A Beginner's Guide to Image Preprocessing Techniques

Intelligent Signal Processing and Data Analysis

SERIES EDITOR
Nilanjan Dey
Department of Information Technology, Techno India College of Technology,
Kolkata, India

Proposals for the series should be sent directly to one of the series editors
above, or submitted to:
Chapman & Hall/CRC
Taylor and Francis Group
3 Park Square, Milton Park
Abingdon, OX14 4RN, UK

https://www.crcpress.com/Intelligent-Signal-Processing-and-Data-
Analysis/book-series/INSPDA

A Beginner's Guide to Image Preprocessing Techniques

Jyotismita Chaki
Nilanjan Dey

CRC Press
Taylor & Francis Group
Boca Raton London New York

CRC Press is an imprint of the
Taylor & Francis Group, an **informa** business

CRC Press
Taylor & Francis Group
6000 Broken Sound Parkway NW, Suite 300
Boca Raton, FL 33487-2742

First issued in paperback 2020

© 2019 by Taylor & Francis Group, LLC
CRC Press is an imprint of Taylor & Francis Group, an Informa business

No claim to original U.S. Government works

ISBN 13: 978-0-367-57080-4 (pbk)
ISBN 13: 978-1-138-33931-6 (hbk)

Library of Congress Cataloging-in-Publication Data

Names: Chaki, Jyotismita, author. | Dey, Nilanjan, 1984- author.
Title: A beginner's guide to image preprocessing techniques / Jyotismita Chaki and Nilanjan Dey.
Description: Boca Raton : Taylor & Francis, a CRC title, part of the Taylor & Francis imprint, a member of the Taylor & Francis Group, the academic division of T&F Informa, plc, 2019. | Series: Intelligent signal processing and data analysis | Includes bibliographical references and index.
Identifiers: LCCN 2018029684| ISBN 9781138339316 (hardback : alk. paper) | ISBN 9780429441134 (ebook)
Subjects: LCSH: Image processing--Digital techniques.
Classification: LCC TA1637 .C7745 2019 | DDC 006.6--dc23
LC record available at https://lccn.loc.gov/2018029684

Visit the Taylor & Francis Web site at
http://www.taylorandfrancis.com

and the CRC Press Web site at
http://www.crcpress.com

Contents

Preface

Digital image processing is a widespread subject and is progressing continuously. The development of digital image processing has been driven by technological improvements in computer processors, digital imaging, and mass storage devices. Digital image processing is used to extract valuable information from images. In this procedure, it additionally deals with (1) enhancement of the quality of an image, (2) image representation, (3) restoration of the original image from its corrupted form, and (4) compression of the bulk amounts of data in the images to increase the efficiency of image retrieval. Digital image processing can be categorized into three different categories. The first category involves the algorithm directly dealing with the raw pixel values like edge detection, image denoising, and so on. The second category involves the algorithm that employs results obtained from the first category for further processing such as edge linking, segmentation, and so forth. The third and last category involves the algorithm that tries to extract semantic information from those delivered by the lower levels such as face recognition, handwriting recognition, and so on. This book covers different image preprocessing techniques, which are essential for the enhancement of image data in order to reduce reluctant falsifications or to improves certain image features vital for additional processing and image retrieval. This book presents the different techniques of image transformation, enhancement, segmentation, morphological techniques, filtering, preprocessing of color images, and preprocessing for Deep Learning in detail. The aim of this book is not only to present different perceptions of digital image preprocessing to undergraduate and postgraduate students, but also to serve as a handbook for practicing engineers. Simulation is an important tool in any engineering field. In this book, the image preprocessing algorithms are simulated using MATLAB®. It has been the attempt of the authors to present detailed examples to demonstrate the various digital image preprocessing techniques.

This book is organized as follows:

- Chapter 1 gives an overview of image preprocessing. The different fundamentals of image preprocessing methods like image correction, image enhancement, image restoration, image compression, and the effect of image preprocessing on image recognition are covered in this chapter. Preprocessing techniques, used to correct the radiometric or geometric aberrations, are introduced in this chapter. The examples related to image correction, image enhancement, image restoration, image compression, and the effect of image preprocessing on image recognition are illustrated through MATLAB examples.

- Chapter 2 deals with pixel brightness transformation techniques. Position-dependent brightness correction is introduced in this chapter. This chapter also gives an overview of different techniques used for grayscale transformation like linear, logarithmic, and power–law or gamma correction. Different types of linear transformations such as identity transformation and negative transformation, different types of logarithmic transformation like log transformations, and inverse log transformations are also included in this chapter. Different image enhancement techniques such as contrast stretching, histogram equalization, and histogram specification are also discussed in this chapter. The examples related to pixel brightness transformation techniques are illustrated through MATLAB examples.

- Chapter 3 is devoted to geometric transformation techniques. Two basic steps in geometric transformations like pixel coordinate transformation or spatial transformation and brightness interpolation are discussed in this chapter. Different simple mapping techniques like translation, scaling, rotation, and shearing are included in this chapter. Also, the affine mapping and different nonlinear mapping techniques such as twirl, ripple, and spherical transformation are discussed step by step. Various brightness interpolation methods like nearest neighbor interpolation, bilinear interpolation, and bicubic interpolation are included in this chapter. The examples related to geometric transformation techniques are illustrated through MATLAB examples.

- Chapter 4 discusses different spatial and frequency filtering techniques. We explain in this chapter different spatial filtering methods such as linear filter, nonlinear filter, and sharpening filter smoothing, which includes smoothing linear filters and order-statistics filters. Various frequency filters like low-pass filter, high-pass filter, bandpass filter, and band-reject filter are also included. In each category of frequency filter, three types of filters are explained: Ideal, Butterworth, and Gaussian. The examples related to different spatial and frequency-filtering techniques are illustrated through MATLAB examples.

- The focus of Chapter 5 is on image segmentation. Different segmentation techniques such as thresholding-based segmentation, edge-based segmentation, and region-based segmentation are explained in this chapter. Different methods to select the threshold value like the histogram shape-based method, entropy-based method, and clustering-based method—which includes k-means and Otsu—are discussed in this chapter. Various edge-based segmentations like Sobel, Canny, Prewitt, Robinson, Robert, kirsch, LoG, and Marr-Hildreth are also explained step by step. Region growing or merging, and region splitting methods are included in region-based segmentation. The examples related to image segmentation techniques are illustrated through MATLAB examples.

- Chapter 6 provides an overview of mathematical morphology techniques. Different methods of binary morphology and grayscale morphology are discussed in this chapter. Binary morphology techniques including erosion, dilation, opening, closing, hit-and-miss, thinning and thickening, as well as grayscale morphology techniques including erosion, dilation, opening and closing are explained. The examples related to mathematical morphology techniques are illustrated through MATLAB examples.

- Chapter 7 deals with preprocessing of color images and preprocessing for neural networks and Deep Learning. Preprocessing of color images includes pseudo color processing, true color processing, different color models, intensity modification, color complement, color slicing, and tone correction. Other types of color image preprocessing involve histogram equalization, segmentation of color image, and so on. Preprocessing for neural networks and Deep Learning includes unvarying aspect ratio, scaling of images, normalization of image inputs, reduction of data dimension, and augmentation of image data. The examples related of preprocessing techniques of color images are illustrated through MATLAB examples.

Dr. Jyotismita Chaki
Jadavpur University

Dr. Nilanjan Dey
Techno India College of Technology

MATLAB® is a registered trademark of The MathWorks, Inc. For product information, please contact:

The MathWorks, Inc.
3 Apple Hill Drive
Natick, MA 01760-2098 USA
Tel: 508 647 7000
Fax: 508-647-7001
E-mail: info@mathworks.com
Web: www.mathworks.com

Authors

Jyotismita Chaki, PhD, was appointed as an Assistant Professor in the School of Computer Engineering at Kalinga Institute of Industrial Technology (KIIT) deemed to be university, India. From 2012 to 2017, she was a Research Fellow at Jadavpur University. Her research interest includes image processing, pattern recognition, computer vision and machine learning. Dr. Chaki is a reviewer of Journal of Visual Communication and Image Representation, Elsevier; Biosystems Engineering, Elsevier; Signal, Image and Video Processing, Springer; Pattern Recognition Letters, Elsevier; Applied Soft Computing, Elsevier and Computers and Electronics in Agriculture, Elsevier.

Nilanjan Dey, PhD, is currently associated with Department of Information Technology, TechnoIndia College of Technology, Kolkata, W.B., India. He holds an honorary position of visiting scientist at Global Biomedical Technologies Inc., California, and is a research scientist at the Laboratory of Applied Mathematical Modeling in Human Physiology, Territorial Organization of Scientific and Engineering Unions, Bulgaria. He is also an associate researcher of Laboratoire RIADI, University of Manouba, TUNISIA. He is an associated member of Wearable Computing Research lab, University of Reading, London, in the United Kingdom.

His research topics are medical imaging, soft computing, data mining, machine learning, rough sets, computer aided diagnosis, and atherosclerosis. He has authored 35 books and 170 journals and 100 international conference papers. He is the editor-in-chief of the *International Journal of Ambient Computing and Intelligence*, US (Scopus, ESCI, ACM dl and DBLP listed), the *International Journal of Rough Sets and Data Analysis*, and is the U.S. co-editor-in-chief of the *International Journal of Synthetic Emotions* and the *International Journal of Natural Computing Research*. He also is the U.S. series editor of *Advances in Geospatial Technologies* (AGT) book series, and the U.S. series editor of *Advances in Ubiquitous Sensing Applications for Healthcare* (AUSAH). He is also the executive editor of the *International Journal of Image Mining* (IJIM) and the associated editor of IEEE Access journal and the *International Journal of Service Science, Management, Engineering and Technology.* He is a life member of IE, UACEE, and ISOC. He has chaired many international conferences such as the ITITS 2017—China, WS4 2017—London, INDIA 2017—Vietnam etc.

1

Perspective of Image Preprocessing on Image Processing

1.1 Introduction to Image Preprocessing

Preprocessing is a typical name for procedures applied to both input and output intensity images. These images are indistinguishable from the original data taken by the sensors. Basically, image preprocessing is a method to transform raw image data into a clean image data, as most of the raw image data contain noise and contain some missing values or incomplete values, inconsistent values, and false values [1]. Missing information means lacking of certain attributes of interest or lacking of attribute values. Inconsistent information means there are some discrepancies in the image. False value means error in the image value. The purpose of preprocessing is an enhancement of the image data to reduce reluctant falsifications or to improve some image features vital for additional processing [2]. Some will contend that image preprocessing is not a smart idea, as it alters or modifies the true nature of the raw data. Nevertheless, smart application of image preprocessing can offer benefits and take care of issues that finally produce improved global and local feature detection [3]. Image preprocessing may have beneficial effects on the excellence of feature extraction and the outcomes of image analysis [4]. Image preprocessing is similar to the scientific standardization of a data set, which is a general step in many feature descriptor techniques. Image preprocessing is used to correct the degradation of an image. In that case, some prior data or information is important such as information about the nature of the degradation, information about the features of the image capturing device, and the conditions under which the image was obtained. Figure 1.1 shows the steps of image preprocessing during digital image processing.

1.2 Complications to Resolve Using Image Preprocessing

The following complications can be resolved by using image preprocessing techniques.

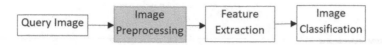

FIGURE 1.1
Image preprocessing step in digital image processing.

1.2.1 Image Correction

Image corrections are generally grouped into radiometric and geometric corrections. Some standard correction methods might be completed prior to the data being delivered to the user [5]. These techniques incorporate a radiometric correction to correct for the irregular sensor reaction over the entire image, and a geometric correction to correct for the geometric misrepresentation owing to different imaging settings such as oblique viewing [6]. Radiometric correction means correcting the radiometric error caused owing to the noise in the brightness values of the image. Some common radiometric errors are random bad pixels or shot noise, line start/stop problems, line or column dropouts, and line or column striping [7]. Radiometric correction is a preprocessing method to rebuild physically aligned values by altering the spectral faults and falsifications caused by the sensors themselves when the individual detectors do not function properly, or are not properly calibrated for the sun's direction and or landscape [8]. For example, shot noise, which is generated when random pixels are not recorded for one or more band, can be corrected by identifying missing pixels. Missing pixel values can be regenerated by taking the average of the neighboring pixels and filling in the value of the missing pixel. Figure 1.2 shows the preprocessed output.

Line start/stop problems occur when scanning detectors fail to start or are out of sequence with other detectors, which results in displaced rows with the pixel data at inappropriate locations along the scan line [9]. This can be solved by determining the rows affected and offsetting and scripting a standard offset for the affected rows. Figure 1.3 shows the preprocessing output.

Line or column dropout error occurs when an entire line does not contain any information and results in blank lines or lines of same gray level value.

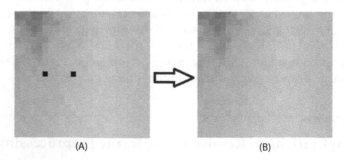

(A)   (B)

FIGURE 1.2
(A) Image with shot noise, (B) Preprocessed output.

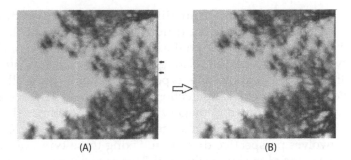

FIGURE 1.3
(A) Image with line start/stop problem, (B) Preprocessed image.

This error can be corrected by averaging above and below pixel values to record in each missing pixels, or fill in values from another image. Figure 1.4 shows the preprocessing output.

Line or column striping occurs when there are some stripes throughout the entire image. This can be resolved by identifying the rows impacted through analysis of a latitudinal geographic profile for the affected band. Figure 1.5 shows the preprocessing output.

FIGURE 1.4
(A) Image with line or column dropout, (B) Preprocessed output.

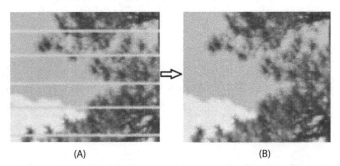

FIGURE 1.5
(A) Image with line striping, (B) Preprocessed output.

Geometric corrections contain modifications of geometric distortions caused by sensor–Earth geometry differences, translation of the data to real-world latitude and longitude on the Earth's surface, motion of platform, Earth curvature, and so on [10]. Geometric correction means putting the pixels in their proper location. This type of correction is generally needed to coregister images for change detection, make accurate distance and area measurements, and correct the imagery distortion. Sometimes the scale of the same object varies due to some change in capturing the image. Geometric error also involves perspective distortion. Fixing these types of distortion involves resampling of the image data [11]. This can be done by determining the correct geometric model, which tells how to transform images, in order to compute the geometric transformations and basically how to analyze the geometric error and resample to produce new output image data. Image row/column coordinates are transformed to real-world coordinates using polynomials [12]. One must choose the proper order of the polynomial. The higher the transformation order, the greater the number of variables in the polynomials and the more the warping stretches and twists in the dataset. The higher order polynomial can provide misleading RMS errors. First order polynomials, called affine transforms, are the linear conversion used to shift the origin of the image, as well as rescale and rotate it. Figure 1.6 shows the outputs of affine transformation.

Second order polynomial is the nonlinear conversion used to correct the camera lens distortion and correct for the Earth's curvature. Figure 1.7 shows the outputs of nonlinear transformation.

1.2.2 Image Enhancement

Image enhancement is mostly refining the sensitivity of information in images for an improved input for automated image processing methods [13]. The primary goal of image enhancement is to adjust image attributes to make them more appropriate for a given task. Through this method, one or more attributes of the image are revised. Image enhancement is used to highlight interesting details in images, remove noise from images, make images more visually appealing, enhance otherwise hidden information, filter

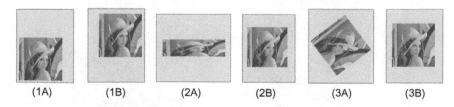

(1A)   (1B)   (2A)   (2B)   (3A)   (3B)

FIGURE 1.6
(1A) Original position of the image, (1B) Shift of origin of the image, (2A) Original scaling of the image, (2B) Rescaling output of the image, (3A) Original orientation of the image, and (3B) Change of orientation of the image.

(1A)   (1B)   (2A)   (2B)

FIGURE 1.7
(1A) Ripple effect, (1B) Nonlinear correction output, (2A) Spherical effect, and (2B) Nonlinear correction output.

important image features, and discard unimportant image features [14]. The enhancement approaches are generally divided into the following two types: spatial domain methods and frequency domain methods. In spatial domain methods, image pixels are enhanced directly. The pixel values are altered to obtain the desired enhancements. The spatial domain image enhancement operation is expressed by using Equation 1.1:

$$S(x,y) = T[I(x,y)], \tag{1.1}$$

where $I(x,y)$ is the input image, $S(x,y)$ is the processed image, and T is an operator on I defined over some neighborhood of (x,y).

Some spatial domain image enhancement includes point processing, mask processing, and so on. In point processing a neighborhood of 1×1 pixel is considered. This is generally used to convert a color image to grayscale or binary image and so forth. In mask processing, the neighborhood is larger than a 1×1 pixel area. This is generally used in image sharpening, image smoothing, and so on. Some of the spatial domain enhancements are shown in Figure 1.8.

With frequency domain methods, first the image is transmitted into the frequency domain. The enhancement procedures are performed on the frequency domain of the image, and then it is again transferred to the spatial domain. Figure 1.9 illustrates the procedure.

The frequency domain image enhancement operation is expressed by using Equation 1.2:

$$F(u,v) = H(u,v)I(u,v), \tag{1.2}$$

where $I(u,v)$ is the input image in the frequency domain, $H(u,v)$ is the transfer function, and $F(u,v)$ is the enhanced image. These enhancement processes are done in order to enhance some frequency parts of the image. Some frequency domain image enhancements are shown in Figure 1.10.

Through image enhancement, the pixel intensity values of the input image are altered according to the enhancement function applied to the input values.

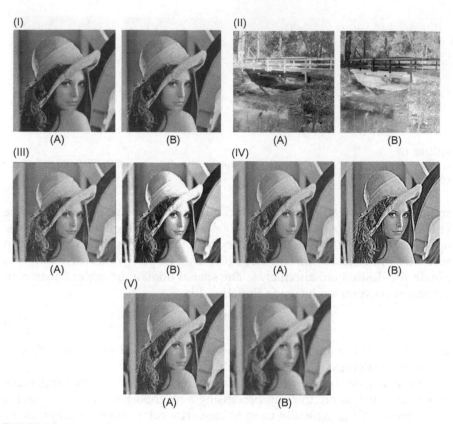

FIGURE 1.8
Enhancement outputs in the spatial domain. (I) Conversion of true color image to grayscale image: (A) True color image, (B) Grayscale Image; (II) Negative Transformation of a grayscale image: (A) Original image, (B) Negative transformed image; (III) Contrast Enhancement: (A) Original image, (B) Contrast enhanced image; (IV) Sharpening an image: (A) Original Image, (B) Sharpen Image; (V) Smoothing an image: (A) Original image, (B) Smoothed image.

1.2.3 Image Restoration

The goal of image restoration methods is to decrease the noise effect or corruption from the image and improve resolution loss. Image preprocessing methods are done both in the image domain and the frequency domain. Corruption may arise in many ways such as noise, motion blur, camera misfocus, and so on [15]. Image restoration is not the same as image enhancement, as the latter one is used to highlight features of the image used

FIGURE 1.9
Steps of enhancement of images in the frequency domain.

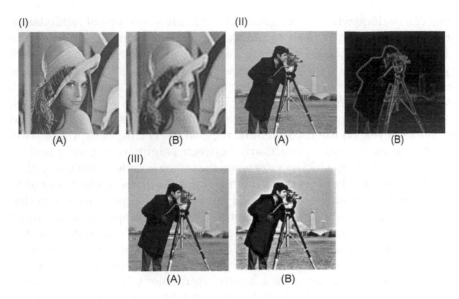

FIGURE 1.10
Enhancement outputs in the frequency domain. (I) The output of low pass filter: (A) Original image, (B) Filtered image; (II) The output of high pass filter: (A) Original image, (B) Filtered image; (III) The output of bandpass filter: (A) Original image, (B) Filtered image.

to make it more attractive to the viewer, but it is not essential in obtaining representative data from a scientific point of view. With image enhancement noise can be efficiently suppressed by losing some resolution, but this is not satisfactory in many applications. Image restoration is useful in these cases. Distorted pixels can be restored by the average value of the neighboring pixels. Some outputs of the restored images are shown in Figure 1.11.

1.2.4 Image Compression

Image compression can be described as the procedure of encoding data using a method that decreases the overall size of the image [16]. This reduction of

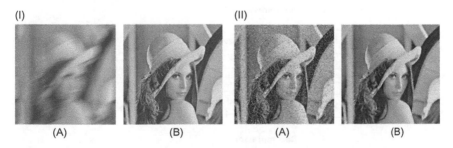

FIGURE 1.11
Outputs of restored images. (I) Output of restoration of blurred image: (A) Blurred image, (B) Restored image; (II) Noise reduction: (A) Noisy image, (B) Image after noise removal.

data can be done when the original dataset holds some type of redundancy. Image compression is used to reduce the total number of bits needed to characterize an image. This can be accomplished by removing different types of redundancy that occur in the pixel values. Generally, three basic redundancies occur in digital images: (1) psycho-visual redundancy, which corresponds to different intensities from image signals sensed by human eyes. Therefore, removing some less important intensities may not be sensed by human eyes; (2) interpixel redundancy, which corresponds to statistical reliance among pixels, particularly between neighboring pixels; and (3) coding redundancy, which occurs when the image is coded with every pixel by a fixed length. There are many methods to deal with these aforementioned redundancies. Compression methods can be classified into two categories: lossy compression and lossless compression. Lossy compression can attain high compression ratios such as 60:1 or higher as it permits some tolerable degradation, but lossless compression can attain very low compression ratios such as 2:1 as it can completely recover the original data. In applications where the image quality is the ultimate requirement, lossless compression is used—such as in medical applications in which no degradation on the original image data are permitted owing to the accuracy requirements for diagnosis. Figure 1.12 shows the block diagram of lossy compression.

Lossy compression is basically a three-stage compression technique to remove the three types of redundancies discussed above. First, a transformation is applied to remove the interpixel redundancy to pack information effectively. Then quantization is applied to eliminate psycho-visual redundancy to characterize the packed information with the fewest bits. The quantized bits are then proficiently encoded to get much more compression from the coding redundancy. Lossy decompression is a perfect inverse technique of lossy compression.

Figure 1.13 shows the block diagram of lossless compression.

Lossless compression is usually a two-step compression technique. First, transformation is applied to the original image to convert it to some other format to reduce the interpixel redundancy. Then an entropy encoder is used

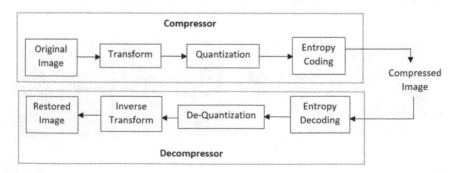

FIGURE 1.12
Block diagram of lossy compression.

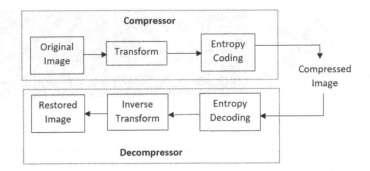

FIGURE 1.13
Block diagram of lossless compression.

to eliminate the coding redundancy. Lossless decompression is a perfect inverse technique of lossless compression.

1.3 Effect of Image Preprocessing on Image Recognition

Image preprocessing is used to enhance the image data so that useful features can be extracted for image recognition. Image cropping is used to crop the irrelevant parts from the image so that the region of interest of the image is focused. Image morphological operations can be applied in some cases. Image filtering is used to create new intensity values in the output image. Smoothing methods are used to remove noise or other small irrelevant data in the image [17]. Filters are also used to highlight the edges of an image. Brightness and contrast of the image can also be adjusted to enhance the useful features of the image [18]. The unwanted areas can be removed from the binary image by using a polygonal mask. Images can also be transformed to different color modes for extraction of different types of features. If the whole scene is rotated, or the image is taken from the wrong perspective, it is required to correct the geometry prior to feature extraction, as many features are dependent on geometric variation [19]. Figure 1.14 shows that

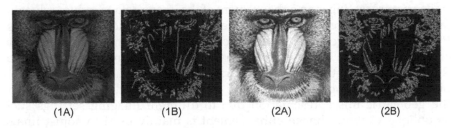

FIGURE 1.14
(1A) Raw image, (1B) Extracted edge information from raw image, (2A) Preprocessed image, and (2B) Extracted edge information from preprocessed image.

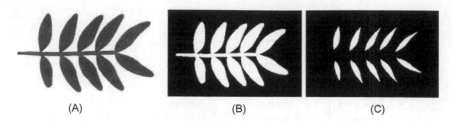

(A)   (B)   (C)

FIGURE 1.15

(A) Original gray image, (B) Binarized image, and (C) Separation of leaflets using morphological erosion operation.

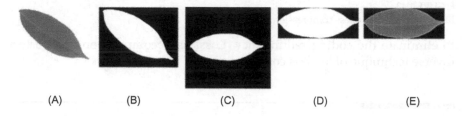

(A)   (B)   (C)   (D)   (E)

FIGURE 1.16

(A) The original image, (B) Binarized image, (C) Corrected orientation, (D) Corrected translation factor, and (E) Preprocessed image after correcting the orientation and translation.

more edge information can be obtained from a preprocessed image than from a raw image.

Suppose we have to count the number of leaflets from a compound leaf image [20]. In this particular example, the preprocessing steps involve binarization and some morphological operations. Figure 1.15 illustrates this. To correct the orientation and translation factor, preprocessing can be applied as shown in Figure 1.16.

1.4 Summary

Image preprocessing is an enhancement of the image data to reduce reluctant falsifications or improve some image features vital for additional processing. Preprocessing is generally used to correct the radiometric or geometric errors, enhance the image, restore the image, and compress the image data. Radiometric correction is used to correct for the irregular sensor reaction over the entire image, and geometric correction is used to compensate for the geometric misrepresentation due to different imaging settings such as oblique viewing. Image enhancement is mainly used to adjust image attributes to make it more appropriate for a given task. The goal of image restoration methods is to decrease the noise effect or corruption from the image and improve resolution loss. Image compression is used to decrease the overall size of the image. Image preprocessing is used to enhance the image data so that useful features can be extracted for image recognition.

References

1. Chatterjee, S., Ghosh, S., Dawn, S., Hore, S., & Dey, N. 2016. Forest type classification: A hybrid NN-GA model based approach. In *Information Systems Design and Intelligent Applications* (pp. 227–236). Springer, New Delhi.
2. Santosh, K. C., & Nattee, C. 2009. A comprehensive survey on on-line handwriting recognition technology and its real application to the Nepalese natural handwriting. *Kathmandu University Journal of Science, Engineering, and Technology*, 5(I), 31–55.
3. Hore, S., Chakroborty, S., Ashour, A. S., Dey, N., Ashour, A. S., Sifaki-Pistolla, D., Bhattacharya, T., & Chaudhuri, S. R. 2015. Finding contours of hippocampus brain cell using microscopic image analysis. *Journal of Advanced Microscopy Research*, 10(2), 93–103.
4. Dey, N., Roy, A. B., Das, P., Das, A., & Chaudhuri, S. S. 2012, November. Detection and measurement of arc of lumen calcification from intravascular ultrasound using harris corner detection. In *Computing and Communication Systems (NCCCS), 2012 National Conference on* (pp. 1–6). IEEE.
5. Santosh, K. C., Lamiroy, B., & Wendling, L. 2012. Symbol recognition using spatial relations. *Pattern Recognition Letters*, 33(3), 331–341.
6. Dey, N., Ahmed, S. S., Chakraborty, S., Maji, P., Das, A., & Chaudhuri, S. S. 2017. Effect of trigonometric functions-based watermarking on blood vessel extraction: An application in ophthalmology imaging. *International Journal of Embedded Systems*, 9(1), 90–100.
7. Saha, M., Chaki, J., & Parekh, R. 2013. Fingerprint recognition using texture features. *International Journal of Science and Research*, 2, 12.
8. Chakraborty, S., Mukherjee, A., Chatterjee, D., Maji, P., Acharjee, S., & Dey, N. 2014, December. A semi-automated system for optic nerve head segmentation in digital retinal images. In *Information Technology (ICIT), 2014 International Conference on* (pp. 112–117). IEEE.
9. Hossain, K., Chaki, J., & Parekh, R. 2014. Translation and retrieval of image information to and from sound. *International Journal of Computer Applications*, 97(21), 24–29.
10. Dey, N., Roy, A. B., & Das, A. 2012, August. Detection and measurement of bimalleolar fractures using Harris corner. In *Proceedings of the International Conference on Advances in Computing, Communications and Informatics* (pp. 45–51). ACM, Chennai, India.
11. Belaïd, A., Santosh, K. C., & d'Andecy, V. P. 2013. Handwritten and printed text separation in real document. *arXiv preprint arXiv:1303.4614*.
12. Dey, N., Nandi, P., Barman, N., Das, D., & Chakraborty, S. 2012. A comparative study between Moravec and Harris corner detection of noisy images using adaptive wavelet thresholding technique. *arXiv preprint arXiv:1209.1558*.

13. Russ, J. C. 2016. *The Image Processing handbook*. CRC Press, Boca Raton, FL.
14. Araki, T., Ikeda, N., Dey, N., Acharjee, S., Molinari, F., Saba, L., Godia, E., Nicolaides, A., & Suri, J. S. 2015. Shape-based approach for coronary calcium lesion volume measurement on intravascular ultrasound imaging and its association with carotid intima-media thickness. *Journal of Ultrasound in Medicine*, 34(3), 469–482.
15. Ashour, A. S., Samanta, S., Dey, N., Kausar, N., Abdessalemkaraa, W. B., & Hassanien, A. E. 2015. Computed tomography image enhancement using cuckoo search: A log transform based approach. *Journal of Signal and Information Processing*, 6(03), 244.
16. Nandi, D., Ashour, A. S., Samanta, S., Chakraborty, S., Salem, M. A., & Dey, N. 2015. Principal component analysis in medical image processing: A study. *International Journal of Image Mining*, 1(1), 65–86.
17. Hangarge, M., Santosh, K. C., Doddamani, S., & Pardeshi, R. 2013. Statistical texture features based handwritten and printed text classification in south indian documents. *arXiv preprint arXiv:1303.3087*.
18. Chaki, J., Parekh, R., & Bhattacharya, S. 2016. Plant leaf recognition using ridge filter and curvelet transform with neuro-fuzzy classifier. In *Proceedings of 3rd International Conference on Advanced Computing, Networking and Informatics* (pp. 37–44). Springer, New Delhi.
19. Chaki, J., Parekh, R., & Bhattacharya, S. 2015. Plant leaf recognition using texture and shape features with neural classifiers. *Pattern Recognition Letters*, 58, 61–68.
20. Chaki, J., Parekh, R., & Bhattacharya, S. In press. Plant leaf classification using multiple descriptors: A hierarchical approach. *Journal of King Saud University-Computer and Information Sciences*, doi:10.1016/j.jksuci.2018.01.007.

2

Pixel Brightness Transformation Techniques

Pixel brightness can be revised by using pixel brightness techniques. The transformation relies on the characteristics of a pixel itself. There are two types of pixel brightness transformations: Position-dependent brightness correction and grayscale transformation [1]. The position-dependent brightness correction, or simply brightness correction, modifies the pixel brightness value by considering the original brightness of the pixel and its position in the image. Grayscale transformation modifies the brightness of the pixel regardless of the position of the pixel.

2.1 Position-Dependent Brightness Correction

The quality of acquisition and digitization of an image is not dependent on the pixel position in the image, however, in many practical cases this theory is not valid [2]. There are different reasons for the degradation of the image quality: first, the uneven sensitivity of the light sensors such as CCD camera elements, vacuum-tube cameras, and so on; second, the nonhomogeneous property of the optical system, that is, if the lens passes farther from the optical lens, the lens decreases light more; and last, the uneven object illumination. With brightness correction, the systematic degradation can be suppressed. The ideal identity transfer function can be described by a multiplicative error coefficient $E(p, q)$. Let the original undegraded, or desired, image be $G(p, q)$ and the image containing degradation $F(p, q)$

$$F(p, q) = E(p, q)G(p, q). \qquad (2.1)$$

If the reference image $G(p, q)$ is captured with a known constant brightness C, then the error coefficient $E(p, q)$ can be obtained. If the image containing degradation is $F_c(p, q)$, the systematic brightness errors can be suppressed by Equation 2.2:

$$G(p, q) = \frac{F(p, q)}{E(p, q)} = \frac{C \times F(p, q)}{F_c(p, q)}. \qquad (2.2)$$

This technique can be adopted if the degradation process of the image is stable. Periodic calibration is needed for the device to find the error coefficient $E(p, q)$.

This method indirectly adopts linearity of the transformation [3]. But as the brightness scale is restricted to some interval, this technique is not true in reality. Equation 2.1 can overflow. This indicates that the best reference image has brightness that is far from both the minimum and maximum limits of the brightness. If the gray scale has 256 levels of brightness, then the best image has a persistent brightness level of 128. Most TV cameras permit us to control the varying illumination settings, as they have automatic gain controllers. This automatic gain control should be switched off first if systematic errors are suppressed using error coefficients.

2.2 Grayscale Transformations

This transformation is not dependent on the pixel position in the image [4]. Here, an input image I is transformed into G by using $\mathcal{T}$. Where $\mathcal{T}$ is the transformation. Let the pixel value of I and G be represented as P_I and P_G, respectively. So, the pixel values are related by Equation 2.3:

$$P_G = \mathcal{T}P_I. \tag{2.3}$$

Using Equation 2.3, the pixel value P_I is mapped to P_G by the transformation function $\mathcal{T}$. As we are dealing only with the grayscale transformation, the output of this transformation is mapped to a grayscale range. The output is mapped to the range $[0, L - 1]$, where $L = 2^m$, m is the number of bits in the image. For example, the range of pixel values of an 8-bit image will be $[0, 255]$.

The following are three basic gray level transformations used in image enhancement:

- Linear transformation
- Logarithmic transformation
- Powerlaw transformation.

Figure 2.1 shows the plots of different grayscale transformation functions.

Here, L represents the number of gray levels. The identity and negative transformation function plots are the types of linear transformations; log and inverse log transformation function plots are the types of logarithmic transformation plots, and nth root and nth power transformation function plots are the types of power-law transformations.

2.2.1 Linear Transformation

There are two types of linear transformation: identity transformation and negative transformation [5]. In identity transformation each pixel value of the input image is directly mapped to the pixel value of the output image. So

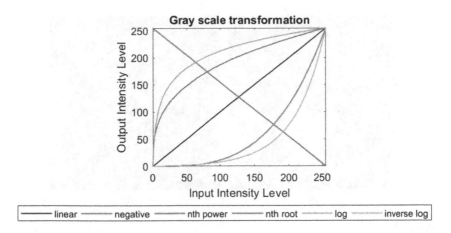

FIGURE 2.1
Different grayscale transformation functions.

the result is same in the input and output image. Hence, it is called identity transformation. The graph of identity transformation is shown in Figure 2.2.

This particular graph shows that between the input and output image there is a straight transition line. This represents that for each input pixel value, the output pixel value will remain the same. So, here the output image is the replica of the input image. Linear transformation can be used to convert a color image into gray scale. Let $I(p, q)$ be the input image and $G(p, q)$ the output image. Then the linear transformation can be represented by Equation 2.4:

$$G(p, q) = I(p, q). \tag{2.4}$$

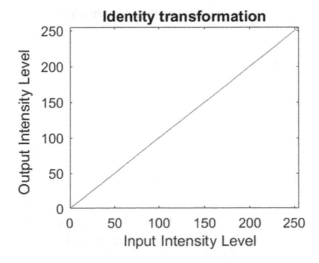

FIGURE 2.2
Identity transformation plot.

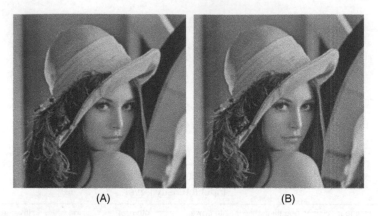

(A)   (B)

FIGURE 2.3
(A) Input image, (B) Linear transformed image.

Figure 2.3 shows the input and output of linear transformation.

The second type of linear transformation is negative transformation [6]. This is basically the inverse of the linear transformation. The negative transformation of an image with gray levels within the range [0, $L - 1$] can be obtained by subtracting each input pixel value from [$L - 1$] and mapping it into the output image, which can be expressed by the Equation 2.5:

$$P_G = L - 1 - P_l. \tag{2.5}$$

This expression indicates the reversing of the gray level intensities of the input pixels, therefore producing a negative image. The graph of negative transformation is shown in Figure 2.4.

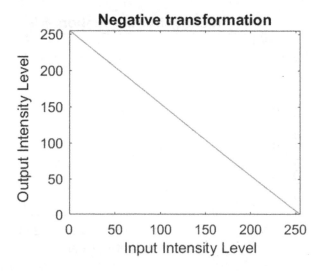

FIGURE 2.4
Negative transformation plot.

Original Image   Negative Transform

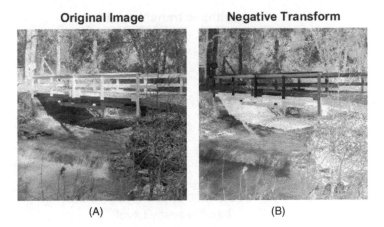

(A)   (B)

FIGURE 2.5
(A) Input image, (B) Negative transformed image.

This technique is beneficial for improving gray or white details implanted in the dark regions of an image. Figure 2.5 shows the input and output of negative transformation.

In the above example, the input image is an 8-bit image. So, there are 256 levels of variations of gray. Putting this in Equation 2.6, we get

$$P_G = 256 - 1 - P_l = 255 - P_l. \tag{2.6}$$

So, by applying the negative transformation, lighter pixels become dark and darker pixels become light.

2.2.2 Logarithmic Transformation

This transformation can be used to brighten the intensity of a pixel in an image [7]. There are various reasons to work with logarithmic intensities rather than with the actual pixel intensity: the logged intensity values are comparatively less dependent on the magnitude of the pixel values, the skewness of the highly skewed values reduces while considering the logs, and the variance estimation increases when using logarithmic values. The visual inspection of data becomes easier by using logged intensities. The raw data are frequently severely clomped together at low intensities followed by a very long tail. Over 75% of the image information may lie in the least 10% values of intensities. The details of such parts are difficult to recognize. After the logarithmic transformation, the change of intensity information is spread out more equally making it simpler to analyze. There are two types of logarithmic transformation: log transformation and inverse log transformation. The graph for log and inverse log transformation is shown in Figure 2.6.

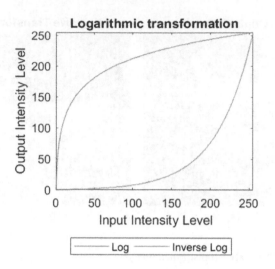

FIGURE 2.6
Logarithmic transformation plot.

The log transformation is used to brighten or increase the detail of the lower intensity values of an image. This can be expressed by the Equation 2.7:

$$P_G = c \times \log(P_I + 1), \tag{2.7}$$

where c is a constant which is normally used to scale the scope of the log transformation function to match the input area. For a 8-bit image, c = 255/log(1 + 255). It can be used to additionally increase the contrast—the higher the c, the brighter the image will appear.

The value 1 is added to every one of the pixel values of the input image in light of the fact that if there is a pixel intensity of 0 in the image, at that point log (0) is equivalent to infinity. So 1 is included, to make the minimum value no less than 1. During log transformation, the dark pixels in an image are extended compared to the higher pixel values. The higher pixel values are somewhat compressed in log transformation. This makes for the improvement of the image.

Figure 2.7 demonstrates the outcomes of log transformation of the original image. We can see that when c = 4, the image is the brightest and the

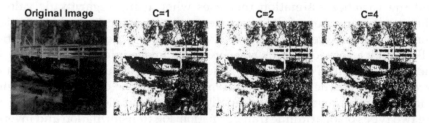

FIGURE 2.7
Results of log transformation.

outspread lines are visible within the tree. These lines are not visible in the original image, as there isn't sufficient contrast in the lower intensities.

Inverse log transformation is opposite of the log transformation. It expands bright regions and compresses the darker intensity level values.

2.2.3 Power-Law Transformation

This transformation is used to increase the contrast of the image [8]. There are two types of power-law transformations: n-th power and n-th root transformation. These transformations can be expressed by Equation 2.8:

$$P_G = C \times P_I^{\gamma}. \tag{2.8}$$

The symbol γ is called gamma and this transformation is also called gamma correction. For different values of γ, various levels of enhancement of the image can be obtained. The graph of power-law transformation is shown in Figure 2.8.

Different monitors or display devices have their own gamma correction. That is the reason they display their image at various intensity. This sort of transformation is used for improving images for various kinds of monitors. The gamma of various monitors is different. For instance, the gamma of monitors lies between 1.8 and 2.5, which implies the image displayed on monitor is dark. The same image with different γ values is shown in Figure 2.9.

Digital images have a finite number of gray levels [9]. Thus, grayscale transformations should be possible using look-up tables. Grayscale transformations are mostly used if the outcome is seen by a human. One way to improve the contrast of the image is contrast stretching (also known as normalization) [10]. Contrast stretching is a linear normalization that expands an arbitrary interval of the intensities of an image and fits this

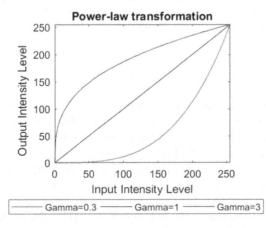

FIGURE 2.8
Power-law transformation plot.

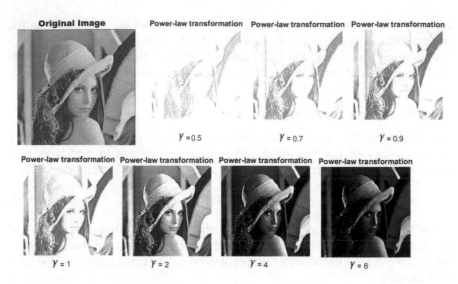

FIGURE 2.9
Results of Gamma variation where $C = 2$.

interval to another arbitrary interval. The initial step is to decide the limits over which image intensity values will be expanded. These lower and upper limits will be known as p and q, individually. For standard 8-bit grayscale images, these limits are normally 0 and 255. Next, the histogram of the input or original image is examined to decide the possible value limits (lower $= a$, upper $= b$) in the unmodified image. If the input image covers the entire possible set of values, direct contrast stretching will achieve nothing, but, even then sometimes the majority of the picture information is contained within a restricted range. This restricted range can be extended linearly with original values, which lie outside the range, being set to the appropriate limit of the extended output range. Then for every pixel, the original value P_I is mapped to output P_G by using Equation 2.9:

$$P_G = (P_I - a)\left(\frac{q-p}{b-a}\right) + p. \tag{2.9}$$

Figure 2.10 shows the result after contrast stretching. In contrast stretching, there exists a one-to-one relationship of the intensity values between the original or input image and the output image; that is, after contrast stretching the input image can be restored from the output image.

Another transformation for contrast improvement is usually applied automatically using histogram equalization, which is a nonlinear normalization, expanding the range of the histogram with high intensities and compressing the areas with low intensities [11]. The point is to discover an image with equally distributed brightness levels over the whole brightness scale. Histogram equalization improves contrast for brightness values close

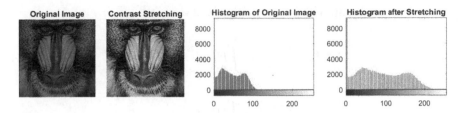

FIGURE 2.10
Contrast stretching results.

to histogram maxima, and decreases contrast near the minima. Figure 2.11 shows the result after histogram equalization. Once histogram equalization is executed, there is no technique for getting back the original image.

Let the input histogram be denoted by H_p where $p_0 \leq p \leq p_t$. The intention is to find a monotonic transform of grayscale $q = \mathcal{T}(p)$, for which the output histogram G_q will remain uniform for the whole input brightness domain, where $q_0 \leq q \leq q_t$. This monotonic property of $\mathcal{T}$ can be expressed by Equation 2.10:

$$\sum_{k=0}^{t} G_{qk} = \sum_{k=0}^{t} H_{pk}.\qquad(2.10)$$

The equalized histogram G_{qk} corresponds to a uniform distribution function $\mathcal{F}$ whose value is constant and can be expressed by Equation 2.11 for a $N \times N$ image,

$$\mathcal{F} = \frac{N^2}{q_t - q_0}.\qquad(2.11)$$

In the continuous case, the ideal continuous histogram is available and can be expressed by Equation 2.12:

$$\int_{q_0}^{q} G(s)ds = \int_{p_0}^{p} H(s)ds.\qquad(2.12)$$

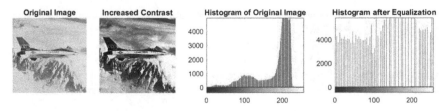

FIGURE 2.11
Histogram equalization result.

Substituting Equation 2.11 in Equation 2.12 we get

$$N^2 \int_{q_0}^{q} \frac{1}{q_t - q_0} \, ds = \int_{p_0}^{p} H(s) ds$$

$$\frac{N^2(q - q_0)}{q_t - q_0} = \int_{p_0}^{p} H(s) ds \tag{2.13}$$

$$q = T(p) = \frac{q_t - q_0}{N^2} \int_{p_0}^{p} H(s) ds + q_0.$$

For discrete case, this is called cumulative histogram, which is approximated by the sum in the digital images and can be expressed by Equation 2.14:

$$q = T(p) = \frac{q_t - q_0}{N^2} \sum_{k=p_0}^{p} H(k) + q_0. \tag{2.14}$$

Histogram specification, or histogram matching, can also be used to enhance the contrast of an image [12]. Histogram specification, or histogram matching, is a method that changes the histogram of one image into the histogram of another image. This change can be effortlessly done by perceiving that if as opposed to using an equally separated perfect histogram (as in histogram equalization), one is specified explicitly. By this method, it is possible to impose an arbitrary histogram of an image to another. First, choose the template histogram. This can be done by determining a specific histogram shape, or by calculating the histogram of a target image. Then, the histogram of the image to be transformed is calculated. Afterwards, calculate the cumulative aggregate of the template histogram. Then, calculate the cumulative aggregate of the histogram of the image to be changed. Finally, map pixels from one bin to another bin, as per the guidelines of histogram equalization. The essential rule is that the actual cumulative aggregate cannot be less than the cumulative aggregate of the template image. Figure 2.12 shows the result of histogram specification.

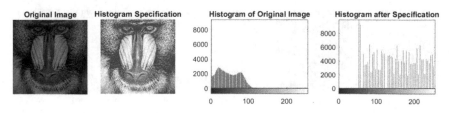

FIGURE 2.12
Result of histogram specification.

2.3 Summary

In image preprocessing, image information captured by sensors on a satellite contains faults associated with geometry and brightness information of the pixels. These errors are improved using suitable techniques. Image enhancement is the adjustment of an image by altering the pixel brightness values to enhance its visual effect. Image enhancement includes a collection of methods used to improve the visual presence of an image, or to alter the image to a form better matched for human or machine understanding. This chapter describes the image enhancement methods by using pixel brightness transformation techniques. Two types of pixel brightness transformation techniques are discussed in this chapter: position dependent and independent, or grayscale transformation. The position-dependent brightness correction modifies the pixel brightness value by considering the original brightness of the pixel and its position in the image. But, grayscale transformation alters the brightness of the pixel regardless the position of the pixel. There are different variations in gray level transformation techniques: linear, logarithmic, and power-law. The identity transformation, which is a type of linear transformation, is mainly used to convert the color image into gray scale. The second type of linear transformation, that is, negative transformation can be used to enhance the gray or white details embedded into the dark region of the image. By using this transformation lighter pixels become dark and darker pixels become light. The logarithmic transformation is used to brighten the intensity of a pixel in an image. The log transformation, which is a type of logarithmic transformation, is used to brighten or increase the detail of the lower intensity values of an image. The second type of logarithmic transformation, that is, inverse log transformation is opposite to the log transformation. Power-law transformation, also known as gamma correction transformation, is used to increase the contrast of an image. For different values of gamma, various levels of enhancement of the image can be obtained. This sort of transformation is used for improving images for various kinds of monitors. To enhance the image contrast, different types of methods can be adopted like contrast stretching, histogram equalization, and histogram specification. Contrast stretching is a linear transformation and the original image can be retrieved from the contrast-stretched image. Histogram equalization is a nonlinear transformation and doesn't allow for the retrieval of the original image from the histogram-equalized image. In case of histogram specification, the histogram of a template image can be applied to the input image to enhance the contrast of the input image.

References

1. Umbaugh, S. E. 2016. *Digital Image Processing and Analysis: Human and Computer Vision Applications with CVIPtools*. CRC Press, Boca Raton, FL.
2. Russ, J. C. 2016. *The Image Processing Handbook*. CRC Press, Boca Raton, FL.

3. Saba, L., Dey, N., Ashour, A. S., Samanta, S., Nath, S. S., Chakraborty, S., Sanches, J., Kumar, D., Marinho, R., & Suri, J. S. 2016. Automated stratification of liver disease in ultrasound: An online accurate feature classification paradigm. *Computer Methods and Programs in Biomedicine*, 130, 118–134.
4. Chaki, J., Parekh, R., & Bhattacharya, S. In press. Plant leaf classification using multiple descriptors: A hierarchical approach. *Journal of King Saud University-Computer and Information Sciences*, doi:10.1016/j.jksuci.2018.01.007.
5. Bhattacharya, T., Dey, N., & Chaudhuri, S. R. 2012. A session based multiple image hiding technique using DWT and DCT. *arXiv preprint arXiv:1208.0950.*
6. Kotyk, T., Ashour, A. S., Chakraborty, S., Dey, N., & Balas, V. E. 2015. Apoptosis analysis in classification paradigm: A neural network based approach. In *Healthy World Conference—A Healthy World for a Happy Life* (pp. 17–22). Kakinada (AP), India.
7. Ashour, A. S., Samanta, S., Dey, N., Kausar, N., Abdessalemkaraa, W. B., & Hassanien, A. E. 2015. Computed tomography image enhancement using cuckoo search: A log transform based approach. *Journal of Signal and Information Processing*, 6(03), 244.
8. Francisco, L., & Campos, C. 2017, October. Learning digital image processing concepts with simple scilab graphical user interfaces. In *European Congress on Computational Methods in Applied Sciences and Engineering* (pp. 548–559). Springer, Cham.
9. Chakraborty, S., Chatterjee, S., Ashour, A. S., Mali, K., & Dey, N. 2018. Intelligent computing in medical imaging: A study. In *Advancements in Applied Metaheuristic Computing* (pp. 143–163). IGI Global, Hershey, Pennsylvania.
10. Negi, S. S., & Bhandari, Y. S. 2014, May. A hybrid approach to image enhancement using contrast stretching on image sharpening and the analysis of various cases arising using histogram. In *Recent Advances and Innovations in Engineering (ICRAIE)*, 2014 (pp. 1–6). IEEE.
11. Dey, N., Roy, A. B., Pal, M., & Das, A. 2012. FCM based blood vessel segmentation method for retinal images. *arXiv preprint arXiv:1209.1181.*
12. Wegner, D., & Repasi, E. 2016, May. Image based performance analysis of thermal imagers. In *Infrared Imaging Systems: Design, Analysis, Modeling, and Testing XXVII* (Vol. 9820, p. 982016). International Society for Optics and Photonics.

3

Geometric Transformation Techniques

Geometric transformations allow the removal of geometric distortion that happens when an image is captured. For example, if one wants to match images of a similar location taken after one year when the later image was perhaps not taken from exactly the same location. To assess changes throughout the year, it is required initially to accomplish a geometric transformation, and afterward, subtract one image from the other. Geometric transformations are often required where the digitized image may be misaligned [1].

There are two basic steps in geometric transformations:

- Pixel coordinate transformation or spatial transformation
- Brightness interpolation.

3.1 Pixel Coordinate Transformation or Spatial Transformation

Pixel coordinate transformation or spatial transformation of an image is a geometric transformation of the image coordinate system, that is, the mapping of one coordinate system onto another. This is characterized by methods of spatial transformation which are mapping functions that builds up a spatial correspondences between every point in the input and output images. Each point in the output adopts the value of its equivalent point in the input image [2]. The correspondence is established via the spatial transformation mapping function to assign the output point onto the input image. It is frequently required to do a spatial transformation to (1) align images captured with different types of sensors or at different times, (2) correct the image distortion caused by the lens and camera orientations, and (3) image morphing or other special effects and so on [3].

An input image comprises known coordinate reference points. The output image consists of the distorted data. The general mapping function can either relate the output coordinate system to that of the input, or vice versa. Let $G(x', y')$ denote the input or original image, and $I(x, y)$ be the deformed (or distorted) image. We can relate corresponding pixels in the two images by Equation 3.1:

$$I \leftrightarrows G. \tag{3.1}$$

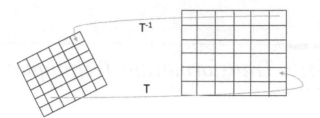

FIGURE 3.1
T: Forward mapping; T^{-1}: Inverse mapping.

Two types of mapping can be done here:

- *Forward Mapping*: Map pixels of input image onto output image, which can be represented by Equation 3.2:

$$G(x', y') = I(x, y). \tag{3.2}$$

- *Inverse Mapping*: Map pixels of output image onto the input image, which can be represented by Equation 3.3:

$$I(x, y) = G(x', y'). \tag{3.3}$$

General mapping example is shown in Figure 3.1.

3.1.1 Simple Mapping Techniques

Translation: Translation means moving the image from one position to another [4]. Let the translation amount in the x and y-direction be t_x and t_y respectively. Translation can be defined by the Equation 3.4:

$$x' = x + t_x$$
$$y' = y + t_y$$

$$\text{OR} \tag{3.4}$$

$$\begin{pmatrix} x' \\ y' \end{pmatrix} = \begin{pmatrix} x \\ y \end{pmatrix} + \begin{pmatrix} t_x \\ t_y \end{pmatrix}.$$

Translation of a geometric shape, as well as an image, is shown in Figure 3.2.

Scaling: Scaling means stretching or contracting an image based on some scaling factors [5,6,7]. Let, s_x and s_y be the scaling factor in the x and y-direction. Scaling can be defined by Equation 3.5:

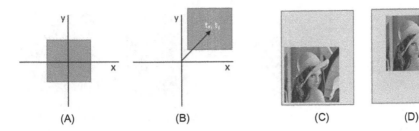

FIGURE 3.2
(A) Original position of the rectangle, (B) Final position of the rectangle after translation, (C) Original position of an image, and (D) Final position of an image after translation.

$$x' = x \cdot s_x$$
$$y' = y \cdot s_y$$

OR (3.5)

$$\begin{pmatrix} x' \\ y' \end{pmatrix} = \begin{pmatrix} s_x & 0 \\ 0 & s_y \end{pmatrix} \cdot \begin{pmatrix} x \\ y \end{pmatrix},$$

$s_x > 1$ represents stretching, $s_x < 1$ represents contracting or shrinking, and $s_x = 1$ means that the size will remain the same.

Scaling of a geometric shape, as well as an image, is shown in Figure 3.3.

Rotation: Rotation means [5,6,7] to change the orientation of an image by an angle of θ, which is defined by Equation 3.6:

$$x' = x \cdot \cos(\theta) - y \cdot \sin(\theta)$$
$$y' = x \cdot \sin(\theta) + y \cdot \cos(\theta)$$

OR (3.6)

$$\begin{pmatrix} x' \\ y' \end{pmatrix} = \begin{pmatrix} \cos(\theta) & -\sin(\theta) \\ \sin(\theta) & \cos(\theta) \end{pmatrix} \cdot \begin{pmatrix} x \\ y \end{pmatrix}.$$

Rotation of a geometric shape as well as of an image is shown in Figure 3.4.

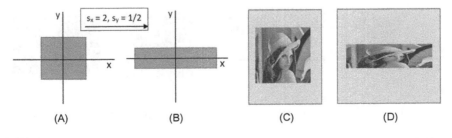

FIGURE 3.3
(A) The original size of the rectangle, (B) Modified size of the rectangle, (C) Original size of the image, and (D) Modified size of the image.

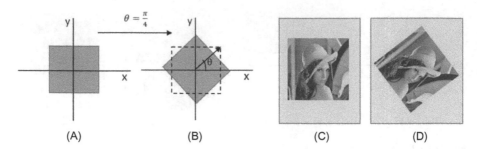

(A)   (B)   (C)   (D)

FIGURE 3.4
(A) Original orientation of rectangle, (B) Modified orientation of rectangle, (C) Original orientation of image, and (D) Modified orientation of image.

Shearing: Images can be sheared along horizontal and vertical direction [8]. For horizontal shears, pixels are relocated horizontally by a distance increasing linearly with the (vertical) distance from the horizontal line, moving to the right above the line and to the left below the line for positive angles. Likewise, for vertical shear, pixels are relocated vertically by a distance that increases linearly with the (horizontal) distance from the vertical line, moving downward to the right of the line, and upward to the left of the line for positive angles. Let, Sh_x and Sh_y be the shear amount in the x and y-direction. Shear can be represented by Equation 3.7:

$$x' = x + Sh_x \cdot y$$
$$y' = y + Sh_y \cdot x$$
$$OR$$

$$\begin{pmatrix} x' \\ y' \end{pmatrix} = \begin{pmatrix} 1 & Sh_x \\ Sh_y & 1 \end{pmatrix} \cdot \begin{pmatrix} x \\ y \end{pmatrix}.$$

(3.7)

Shear of a geometric shape, as well as an image, is shown in Figure 3.5.

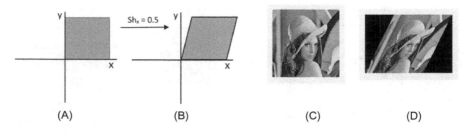

(A)   (B)   (C)   (D)

FIGURE 3.5
(A) Original rectangle, (B) Horizontal sheared rectangle, (C) Original image, and (D) Horizontal sheared image.

3.1.2 Affine Mapping

All possible simple mapping or transformations are special cases of affine mapping [9,10]. The affine transformation is the combination of simple transformations. Affine mapping is a linear mapping method, which conserves straight lines, planes, and points. Sets of parallel lines stay parallel after an affine transformation.

The overall affine transformation is normally written in homogeneous coordinates, as shown in Equation 3.8:

$$\begin{pmatrix} x' \\ y' \end{pmatrix} = P \times \begin{pmatrix} x \\ y \end{pmatrix} + Q. \tag{3.8}$$

By defining only the Q matrix, this transformation turns to a pure translation transformation, as shown in Equation 3.9:

$$P = \begin{pmatrix} 1 & 0 \\ 0 & 1 \end{pmatrix}, \quad Q = \begin{pmatrix} t_x \\ t_y \end{pmatrix}. \tag{3.9}$$

By defining only the P matrix, this transformation turns into a pure rotation transformation (for positive or clockwise rotation), as shown in Equation 3.10:

$$P = \begin{pmatrix} \cos(\theta) & -\sin(\theta) \\ \sin(\theta) & \cos(\theta) \end{pmatrix}, \quad Q = \begin{pmatrix} 0 \\ 0 \end{pmatrix}. \tag{3.10}$$

Similarly, pure scaling can be defined by Equation 3.11:

$$P = \begin{pmatrix} s_x & 0 \\ 0 & s_y \end{pmatrix}, \quad Q = \begin{pmatrix} 0 \\ 0 \end{pmatrix}. \tag{3.11}$$

Since the general affine transformation is characterized by six constants, it is conceivable to express this transformation by determining the new output image locations (x', y') of any three input image coordinate (x, y) pairs. In general, several points are estimated and a least squares technique is used to find the finest fitting transform.

3.1.3 Nonlinear Mapping

Twirl: In case of twirl, rather than using image color at (x', y'), use image colors at twirled (x, y) position [11]. Rotate or turn the image by an angle θ at the anchor point or center (x_c, y_c). Progressively, turn the image as the spiral distance S from the center increases up to $S_{\max}$. The image remains

Original Image   **After Twirl**

FIGURE 3.6
Twirl effect of an image.

unaffected outside of the radial distance $S_{\max}$. Twirl can be defined by Equation 3.12:

$$D_x = x' - x_c, \quad D_y = y' - y_c$$

$$S = \sqrt{D_x^2 + D_y^2}$$

$$\alpha = \arctan(D_y, D_x) + \theta \cdot \left(\frac{S_{\max} - S}{S_{\max}}\right)$$

$$x = \begin{cases} x_c + r \cdot \cos(\alpha) & \text{if } S \leq S_{\max} \\ x' & \text{if } S > S_{\max} \end{cases}$$

$$y = \begin{cases} y_c + r \cdot \sin(\alpha) & \text{if } S \leq S_{\max} \\ y' & \text{if } S > S_{\max}. \end{cases}$$

(3.12)

The twirl effect of an image is shown in Figure 3.6.

Ripple: Ripple effects are like wave patterns, which are introduced in the image along both the x and y-directions [12]. Let the amplitude of the wave pattern in the x and y-direction is defined as A_x and A_y, respectively, and the frequency of the wave in the x and y-direction is defined as F_x and F_y, respectively. So, this effect can be expressed by the sinusoidal function, as shown in Equation 3.13:

$$x = x' + A_x \cdot \sin\left(\frac{2\pi \cdot y'}{F_x}\right)$$

$$y = y' + A_y \cdot \sin\left(\frac{2\pi \cdot x'}{F_y}\right).$$

(3.13)

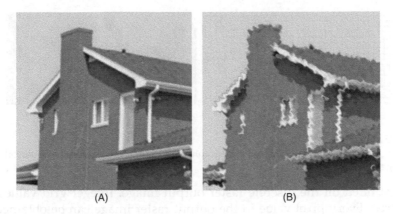

(A)   (B)

FIGURE 3.7
(A) Original image, (B) Ripple effect.

The ripple effect of an image is shown in Figure 3.7.

Spherical Transformation: This transformation zooms in the center of the image. Let the center of the lens be (x_c, y_c), L_{max} the lens radius, and τ is the refraction index [13]. The spherical transformation is defined by Equation 3.14:

$$D_x = x' - x_c, \quad D_y = y' - y_c,$$
$$S = \sqrt{D_x^2 + D_y^2}$$
$$Z = \sqrt{L_{max}^2 + S^2}$$
$$\alpha_x = \left(1 - \frac{1}{\tau}\right) \cdot \sin^{-1}\left(\frac{D_x}{\sqrt{D_x^2 + Z^2}}\right)$$
$$\alpha_y = \left(1 - \frac{1}{\tau}\right) \cdot \sin^{-1}\left(\frac{D_y}{\sqrt{D_y^2 + Z^2}}\right) \tag{3.14}$$
$$x = x' - \begin{cases} Z \cdot \tan(\alpha_x) & \text{if } S \le L_{max} \\ 0 & \text{if } S > L_{max} \end{cases}$$
$$y = y' - \begin{cases} Z \cdot \tan(\alpha_y) & \text{if } S \le L_{max} \\ 0 & \text{if } S > L_{max}. \end{cases}$$

The spherical transformation effect of an image is shown in Figure 3.8.

3.2 Brightness Interpolation

New pixel coordinates were found after the geometric transformation has been performed [14]. The location of the new coordinate point usually does

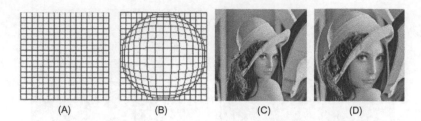

FIGURE 3.8
(A) Original graph, (B) Spherical effect of graph, (C) Original image, and (D) Spherical effect of image.

not get fitted on the discrete raster output image. Integer grid values are required. Every pixel value in the output raster image can be obtained by brightness interpolation of some noninteger neighboring samples. The brightness interpolation is generally done by defining the brightness of the original pixel in the input image that resembles the pixel in the output discrete raster image. Interpolation is used when we need to estimate the value of an unknown pixel by using some known data.

3.2.1 Nearest Neighbor Interpolation

This is the simplest interpolation approach [15]. This technique basically determines the nearest neighboring pixel value and adopts its intensity value, as shown in Figure 3.9.

Consider the following example (Figure 3.10).

Figure 3.9 shows that the 2D input matrix is 3 × 3 and it is interpolated to 6 × 6. First, we must find the ratio of the input and output matrix size, as shown in Equation 3.15:

$$R_{row} = \frac{3}{6}, \quad R_{col} = \frac{3}{6}. \tag{3.15}$$

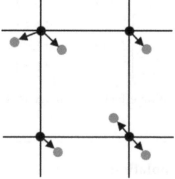

FIGURE 3.9
Black pixels: Original pixels, Red pixels: Interpolated pixels.

$$\begin{bmatrix} 5 & 8 & 10 \\ 6 & 9 & 11 \\ 7 & 4 & 12 \end{bmatrix} \xrightarrow{\text{Nearest Neighbor Interpolation}} \begin{bmatrix} 5 & 5 & 8 & 8 & 10 & 10 \\ 5 & 5 & 8 & 8 & 10 & 10 \\ 6 & 6 & 9 & 9 & 11 & 11 \\ 6 & 6 & 9 & 9 & 11 & 11 \\ 7 & 7 & 4 & 4 & 12 & 12 \\ 7 & 7 & 4 & 4 & 12 & 12 \end{bmatrix}$$

FIGURE 3.10
Nearest neighbor interpolation output.

Then, based on the output matrix size the row-wise and column-wise pixel positions are normalized.

$$\text{Row}_{\text{position}} = \frac{[1 \quad 2 \quad 3 \quad 4 \quad 5 \quad 6]}{R_{\text{row}}} = [0.5 \quad 1 \quad 1.5 \quad 2 \quad 2.5 \quad 3] = \begin{bmatrix} 1 \\ 1 \\ 2 \\ 2 \\ 3 \\ 3 \end{bmatrix}$$

$$\text{Col}_{\text{position}} = \frac{[1 \quad 2 \quad 3 \quad 4 \quad 5 \quad 6]}{R_{\text{col}}} = [0.5 \quad 1 \quad 1.5 \quad 2 \quad 2.5 \quad 3] = [1 \quad 1 \quad 2 \quad 2 \quad 3 \quad 3].$$

$$(3.16)$$

After that, the row-wise interpolation is performed on all columns. The output of the first column after interpolation is shown in Figure 3.11.

The row-wise interpolation output is shown below:

$$\begin{bmatrix} 5 & 8 & 10 \\ 5 & 8 & 10 \\ 6 & 9 & 11 \\ 6 & 9 & 11 \\ 7 & 4 & 12 \\ 7 & 4 & 12 \end{bmatrix}$$

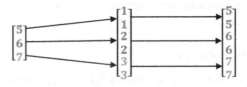

FIGURE 3.11
Row-wise interpolation of the first column of the input matrix.

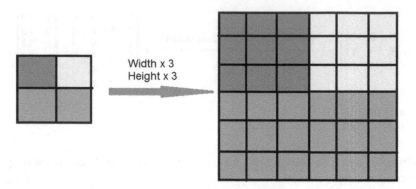

FIGURE 3.12
Nearest neighbor interpolation of an image.

Similarly, the column-wise interpolation for all rows is shown below:

$$\begin{bmatrix} 5 & 5 & 8 & 8 & 10 & 10 \\ 6 & 6 & 9 & 9 & 11 & 11 \\ 7 & 7 & 4 & 4 & 12 & 12 \end{bmatrix}$$

The final nearest neighbor interpolated output matrix is shown below:

$$\begin{bmatrix} 5 & 5 & 8 & 8 & 10 & 10 \\ 5 & 5 & 8 & 8 & 10 & 10 \\ 6 & 6 & 9 & 9 & 11 & 11 \\ 6 & 6 & 9 & 9 & 11 & 11 \\ 7 & 7 & 4 & 4 & 12 & 12 \\ 8 & 7 & 4 & 4 & 12 & 12 \end{bmatrix}$$

The nearest neighbor interpolation output of an image is shown in Figure 3.12.

The position error of the nearest neighborhood interpolation is at most half a pixel. This error is perceptible on objects with straight-line boundaries, which may appear step-like after the transformation.

In nearest neighbor interpolation, each nearby pixel has similar characteristics, hence, it becomes easier to add or remove the pixels as per requirement. The major drawback of this method is unwanted artifacts, like the sharpening of edges that may appear in an image while resizing, hence, it is generally not preferred.

3.2.2 Bilinear Interpolation

This type of interpolation searches four neighboring points of the interpolated point (x, y), as shown in Figure 3.13, and assumes that the brightness function is linear in this neighborhood [16].

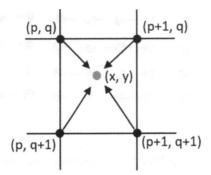

FIGURE 3.13
Black pixels: Original pixels, Red pixel: Interpolated pixel.

Consider a discrete image function. The black circles represents the known pixels of the image I, and the red circle is lying outside the known samples. This interpolation is not linear but the product of two linear functions. If the interpolated point lies on one of the edges of the cell $[(I(p, q) \rightarrow I(p + 1, q)), (I(p + 1, q) \rightarrow I(p + 1, q + 1)), (I(p + 1, q + 1) \rightarrow I(p, q + 1)), (I(p, q + 1) \rightarrow I(p, q))]$, the function becomes linear. Otherwise, the bilinear interpolation function is quadratic. The interpolated value of (x, y) is considered as a linear combination of four known sample values, that is, $I(p, q)$, $I(p + 1,q)$, $I(p + 1, q + 1)$, $I(p, q + 1)$. The influence of each samples depends on the proximity to the interpolated point in the linear combination.

$$I(x,y) = (1-tx)*(1-ty)*I(p,q)+tx*(1-ty)*I(p+1,q)$$
$$+ ty*(1-tx)*I(p,q+1)+tx*ty*I(p+1,q+1), \quad (3.17)$$

where $= x - p$, $ty = y - q$.

A minor reduction in resolution and blurring can happen while using bilinear interpolation due to its averaging nature. The problem of step-like straight boundaries with the nearest neighborhood interpolation is reduced when using bilinear interpolation. The main advantage of using bilinear interpolation is that it is fast and simple to implement.

3.2.3 Bicubic Interpolation

Improves the model of the brightness function by using sixteen neighboring points for interpolation [17]. This interpolation fits a series of cubic polynomials to the brightness values contained in the 4×4 array of pixels surrounding the calculated address. First, interpolation is done along the x-direction using the 16 grid samples (black), as shown in Figure 3.14. Then, interpolation is done along the other dimension (blue line) by using the interpolated pixels from the previous step.

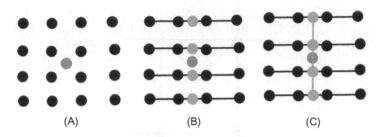

FIGURE 3.14
(A) known pixel (black) and interpolated pixel (red), (B) x-direction interpolation, (C) y-direction interpolation.

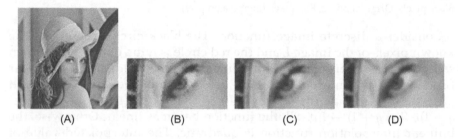

FIGURE 3.15
(A) Original image, (B) Nearest-neighbor interpolation, (C) Bilinear interpolation, and (D) Bicubic interpolation.

Bicubic interpolation does not suffer from the step-like boundary problem of nearest neighborhood interpolation and copes with linear interpolation blurring as well. Bicubic interpolation is often used in raster displays that enable zooming with respect to an arbitrary point—if the nearest neighborhood method were used, areas of the same brightness would increase. Bicubic interpolation preserves fine details in the image very well.

The comparison between nearest-neighbor, bilinear, and bicubic interpolation is shown in Figure 3.15.

3.3 Summary

Geometric transformation is actually the rearrangement of pixels of the image. Coordinates of the input image is transformed into the coordinates of the output image using some transformation function. The output pixel intensity of a specified pixel position may not depend on the pixel intensity of that particular input pixel, but is dependent on the position as specified in the transformation matrix. There are two types of geometric transformation: pixel coordinate transformation and brightness interpolation. Pixel coordinate transformation, or spatial transformation, of an image is a geometric transformation of the image coordinate system, that is, the mapping of one coordinate system onto another. Mapping can be forward (map pixels of an input image onto an output image) or backward (map pixels of an output image onto an input image). This type of transformation involves some linear mapping like translation, scaling, rotation, shearing, and affine transformation. Nonlinear mapping involves twirl, ripple, and spherical transformation. The brightness interpolation is generally done by defining the brightness of the original pixel in the input image that resembles the pixel in the output discrete raster image. Brightness interpolation involves nearest neighbor interpolation, bilinear interpolation, and bicubic interpolation.

References

1. Candemir, S., Borovikov, E., Santosh, K. C., Antani, S., & Thoma, G. 2015. Rsilc: Rotation-and scale-invariant, line-based color-aware descriptor. *Image and Vision Computing*, 42, 1–12.
2. Chaki, J., Parekh, R., & Bhattacharya, S. 2015. Plant leaf recognition using texture and shape features with neural classifiers. *Pattern Recognition Letters*, 58, 61–68.
3. Gonzalez, R. C., Woods, R. E. 2016. Digital Image Processing 3rd edition, Prentice-Hall, New Jersey. ISBN-9789332570320, 9332570329.
4. Chaki, J., Parekh, R., & Bhattacharya, S. In press. Plant leaf classification using multiple descriptors: A hierarchical approach. *Journal of King Saud University-Computer and Information Sciences*, doi:10.1016/j.jksuci.2018.01.007.
5. Li, J., Yu, C., Gupta, B. B., & Ren, X. 2018. Color image watermarking scheme based on quaternion Hadamard transform and Schur decomposition. *Multimedia Tools and Applications*, 77(4), 4545–4561.
6. Chakraborty, S., Chatterjee, S., Ashour, A. S., Mali, K., & Dey, N. 2018. Intelligent computing in medical imaging: A study. In *Advancements in Applied Metaheuristic Computing* (pp. 143–163). IGI Global.
7. Chaki, J., Parekh, R., & Bhattacharya, S. 2016, December. Recognition of plant leaves with major fragmentation. In *Computational Science and Engineering: Proceedings of the International Conference on Computational Science and Engineering* (Beliaghata, Kolkata, India, October 4–6, 2016) (p. 111). CRC Press, Boca Raton, FL.
8. Chaki, J., Parekh, R., & Bhattacharya, S. 2015, July. Recognition of whole and deformed plant leaves using statistical shape features and neuro-fuzzy classifier. In *Recent Trends in Information Systems (ReTIS), 2015 IEEE 2nd International Conference on* (pp. 189–194). IEEE, Kolkata, India.
9. Vučković, V., Arizanović, B., & Le Blond, S. 2018. Ultra-fast basic geometrical transformations on linear image data structure. *Expert Systems with Applications*, 91, 322–346.

10. Santosh, K. C., Lamiroy, B., & Wendling, L. 2011, August. DTW for matching radon features: A pattern recognition and retrieval method. In *International Conference on Advanced Concepts for Intelligent Vision Systems* (pp. 249–260). Springer, Berlin, Heidelberg.
11. Sonka, M., Hlavac, V., & Boyle, R. 2014. *Image Processing, Analysis, and Machine Vision*. Cengage Learning, Stamford, USA.
12. Fu, K. S. 2018. *Special Computer Architectures for Pattern Processing*. CRC Press, Boca Raton, FL.
13. Gilliam, C., & Blu, T. 2018. Local all-pass geometric deformations. *IEEE Transactions on Image Processing*, 27(2), 1010–1025.
14. Chaki, J., Parekh, R., & Bhattacharya, S. 2016. Plant leaf recognition using ridge filter and curvelet transform with neuro-fuzzy classifier. In *Proceedings of 3rd International Conference on Advanced Computing, Networking and Informatics* (pp. 37–44). Springer, New Delhi.
15. Jiang, N., & Wang, L. 2015. Quantum image scaling using nearest neighbor interpolation. *Quantum Information Processing*, 14(5), 1559–1571.
16. Wegner, D., & Repasi, E. 2016, May. Image based performance analysis of thermal imagers. In *Infrared Imaging Systems: Design, Analysis, Modeling, and Testing XXVII* (Vol. 9820, p. 982016). International Society for Optics and Photonics, Baltimore, Maryland, United States.
17. Dong, C., Loy, C. C., He, K., & Tang, X. 2016. Image super-resolution using deep convolutional networks. *IEEE Transactions on Pattern Analysis and Machine Intelligence*, 38(2), 295–307.

4

Filtering Techniques

Filtering is a method for enhancing or altering an image [1]. There are mainly two types of filtering:

- Spatial Filtering
- Frequency Filtering

4.1 Spatial Filter

In spatial filtering, the processed pixel value for the existing pixel is dependent on both itself and neighboring pixels [2]. Therefore, spatial filtering is a neighboring procedure, where the value of any particular pixel in the output image is calculated by applying some algorithm to the values of the neighboring pixels of the corresponding input pixel [3]. A pixel's neighborhood is defined by a set of surrounding pixels relative to that pixel. Some types of spatial filtering are discussed below.

4.1.1 Linear Filter (Convolution)

The result of the linear filtering [4] is the summation of products of the mask coefficients with the equivalent pixels exactly beneath the mask, as shown in Figure 4.1.

Linear filtering can be expressed by Equation 4.1:

$$
\begin{aligned}
I(x,y) = & [M(-1,1) * I(x-1,y+1)] \\
& + [M(0,1) * I(x,y+1)] + [M(1,1) * I(x+1,y+1)] \\
& + [M(-1,0) * I(x-1,y)] + [M(0,0) * I(x,y)] \\
& + [M(1,0) * I(x+1,y)] + [M(-1,-1) * I(x-1,y-1)] \\
& + [M(0,-1) * I(x,y-1)] + [M(1,-1) * I(x+1,y-1)]. \quad (4.1)
\end{aligned}
$$

The mask coefficient $M(0, 0)$ overlaps with image pixel value $I(x, y)$, representing that the mask center is at (x, y) when the calculation of the sum

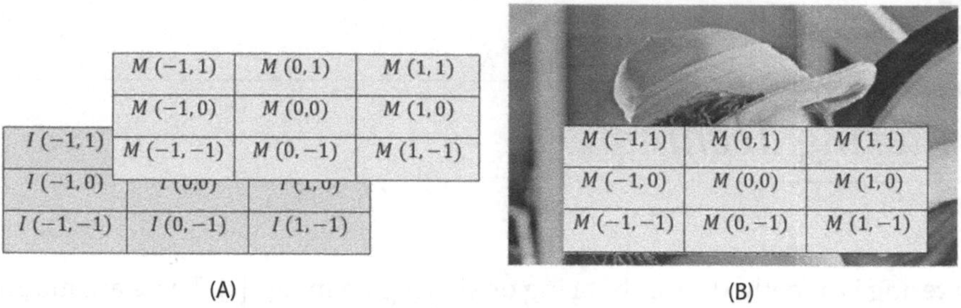

FIGURE 4.1

(A) I: Image pixel positions and M: Mask Coefficients, (B) Mask of image pixels.

of products occurred. For a mask of size $p \times q$, p and q are odd numbers and represented as $p = 2m + 1$, $q = 2n + 1$, where m and n are nonnegative integers. Linear filtering of an image I of size $p \times q$, with a filter mask of size $p \times q$, is given by the Equation 4.2:

$$LF(x, y) = \sum_{a=-m}^{m} \sum_{b=-n}^{n} M(a, b) * I(x + a, y + b). \tag{4.2}$$

4.1.2 Nonlinear Filter

Nonlinear spatial filtering also works on neighborhoods, as discussed in the case of linear filtering [5]. The only difference is the nonlinear filtering is based conditionally on the values of the neighboring pixels of a relative pixel.

4.1.3 Smoothing Filter

Smoothing filters are mainly used to reduce noise of an image and for blurring [6,7]. Blurring is used to remove unimportant information from an image prior to feature extraction, and is used to connect small breaks in curves or lines. Blurring is also used to reduce noise from an image. A smoothing filter is also useful for highlighting gross details. Two types of smoothing spatial filters exist:

- Smoothing Linear Filters
- Order-Statistics Filters

A smoothing linear filter is basically the mean of the neighborhood pixels of the filter mask. Therefore, this filter is sometimes called "mean filter" or "averaging filter." The concept entails substituting the value of every single pixel in an image with the mean of the neighborhood pixels defined by the filter mask. Figure 4.2 shows a 3×3 standard mean and weighted mean smoothing linear filter:

$\frac{1}{9}$ ×	1	1	1
	1	1	1
	1	1	1

$\frac{1}{16}$ ×	1	2	1
	2	4	2
	1	2	1

(A)   (B)

FIGURE 4.2
(A) Standard mean smoothing linear filter, (B) Weighted mean smoothing linear filter.

Filtering an $I(m \times n)$ image with a weighted averaging filter of size $m \times n$ is given by Equation 4.3:

$$SF(x,y) = \frac{\sum_{a=-m}^{m}\sum_{b=-n}^{n} M(a,b)*I(x+a,y+b)}{\sum_{a=-m}^{m}\sum_{b=-n}^{n} M(a,b)}. \qquad (4.3)$$

The output of a smoothing linear filter is shown in Figure 4.3.

Order-statistics smoothing filters are basically nonlinear spatial filter [8–10]. The response of this filter is constructed by ordering or ranking the pixels enclosed in the image area covered by the filter. Then, the value of the center pixel is replaced with the value calculated by the ordering or ranking result. This type of filter is also known as "median filter." The median filter is used to reduce the salt and pepper type noise from an image while preserving edges [11–13]. This filter works by moving a window of a particular size over each and every pixel of the image, and replaces each pixel value with the median of the neighboring pixel values. To calculate the median, first the pixel values beneath the window are sorted into numerical order and then the considered pixel value is replaced with the median pixel value of the sorted list.

Consider the example shown in Figure 4.4. A 3 × 3 window is used in this example.

Figure 4.5 shows the output of a median filter when applied to salt and pepper noise image.

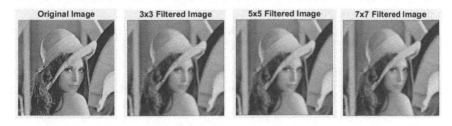

Original Image   3x3 Filtered Image   5x5 Filtered Image   7x7 Filtered Image

FIGURE 4.3
Smoothing linear filter output.

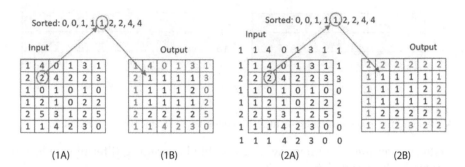

      (1A)   (1B)   (2A)   (2B)

FIGURE 4.4
(1) Keeping the border value unchanged: (1A) Input Image values, (1B) Output after smoothing; (2) Boundary values are also filtered by extending the border values: (2A) Input Image values, (2B) Output after smoothing.

Image with Salt and Pepper Noise   **Median Filtered Image**

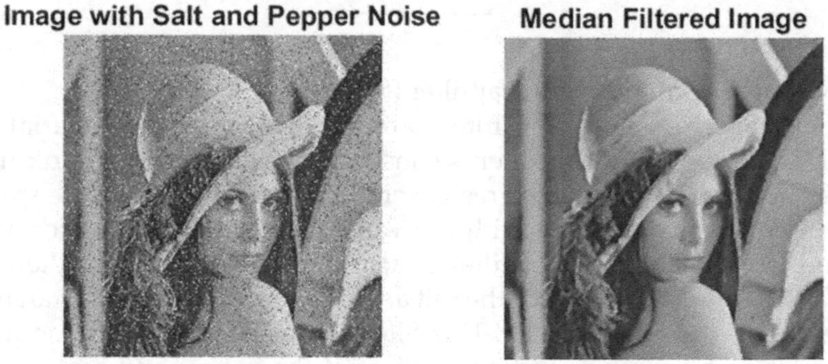

FIGURE 4.5
Median filter output.

4.1.4 Sharpening Filter

The primary goal of this filter is to enhance the fine detail in an image or to highlight the blurred detail [14]. Sharpening can be performed by using spatial derivatives, which can be applied in areas of flat regions or constant gray level regions, at the step and end of discontinuities or ramp discontinuities, and along gray-level discontinuities or ramps. These discontinuities can be lines, noise points, and edges.

The first order partial spatial derivatives of a digital image $I(x, y)$ can be expressed by using Equation 4.4:

$$\frac{\partial I}{\partial x} = I(x+1,y) - I(x,y) \quad \text{and} \quad \frac{\partial I}{\partial y} = I(x,y+1) - I(x,y). \tag{4.4}$$

First order partial derivative must be (1) zero in flat regions, (2) nonzero at the step and gray level ramp discontinuities, and (3) nonzero along ramps.

Original image **Sharpened image**

FIGURE 4.6
Sharpen image output.

The second order partial spatial derivatives of a digital image $I(x, y)$ can be expressed by using Equation 4.5:

$$\frac{\partial^2 I}{\partial x^2} = I(x+1,y) + I(x-1,y) - 2I(x,y)$$

$$\frac{\partial^2 I}{\partial y^2} = I(x,y+1) + I(x,y-1) - 2I(x,y). \tag{4.5}$$

Second order partial derivative must be: (1) zero in flat regions, (2) nonzero at the step and gray level ramp discontinuities, (3) zero along ramps of constant slope.

The first order derivative is nonzero along the entire discontinuity or ramp, but the second order derivative is nonzero only at the step and gray level ramp discontinuities. A first order derivative is used to make the edge thick, and a second-order derivative is used to enhance or highlight fine details such as thin edges and lines, including noise.

Figure 4.6 shows the result of a sharpening filter.

4.2 Frequency Filter

Frequency filters are used to process an image in the frequency domain [15]. The image is converted to frequency domain by using a Fourier transform function. After frequency domain processing, the image is retransformed into the spatial domain by inverse Fourier transform. Reducing high frequencies in the spatial domain converts the image into a smoother one, while reducing low frequencies highlights the edges of the image [16]. All frequency filters can also be implemented in the spatial domain, and frequency filters are computationally not costly to accomplish filtering in the spatial domain. Frequency filtering is also more suitable if there is no direct kernel that can be created in the spatial domain, in which case they may also be more effective. All spatial domain images have an equivalent frequency representation. The high frequency corresponds to pixel values that rapidly vary across the image like leaves, text, texture, and so forth. Low frequency corresponds to the homogeneous part of the image.

Frequency filtering is founded on the Fourier Transform. The operator generally takes a filter function and an image in the Fourier domain. This image is then multiplied in a pixel-by-pixel fashion with the filter function, and can be expressed by Equation 4.6:

$$F(u,v) = \frac{1}{PQ} \sum_{x=0}^{P-1} \sum_{y=0}^{Q-1} I(x,y) e^{-j2\pi\left(\frac{ux}{P} + \frac{vy}{Q}\right)}. \tag{4.6}$$

Here $I(x, y)$ is the input image of dimension $P \times Q$ in the Fourier domain and $F(u, v)$ is the filtered image [$u = 0, \dots, P - 1$ and $v = 0, \dots, Q - 1$]. To convert the frequency domain image into the spatial domain, $F(u,v)$ is retransformed by using the inverse Fourier Transform, as shown in Equation 4.7:

$$I(x,y) = \frac{1}{PQ} \sum_{x=0}^{P-1} \sum_{y=0}^{Q-1} F(u,v) e^{-j2\pi\left(\frac{ux}{P} + \frac{vy}{Q}\right)}. \tag{4.7}$$

Since the multiplication in the Fourier space is identical to convolution in the spatial domain, all frequency filters can be implemented theoretically as a spatial filter. Different types of frequency filters are discussed in the following subsections.

4.2.1 Low-Pass Filter

A low-pass filter is a filter that passes or allows low-frequency signals, and suppresses signals with higher frequencies than the cutoff or threshold frequency [17]. Based on the specific filter design, the actual amount of suppression varies for each frequency. A low-pass filter is generally used to smooth an image. The standard forms of low-pass filters are Ideal, Butterworth, and Gaussian low-pass filters.

4.2.1.1 Ideal Low-Pass Filter (ILP)

This is the simplest low-pass filter that suppresses all high-frequency components of the Fourier Transform that are greater than a specified

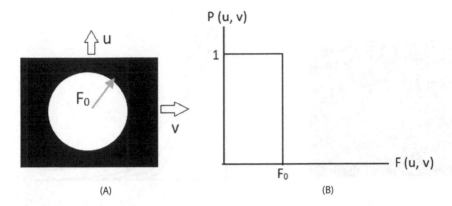

FIGURE 4.7
(A) Filter displayed as an image, (B) Graphical representation of ideal low-pass filter.

cutoff frequency F_0. This transfer function of the filter can be defined by Equation 4.8:

$$P(u,v) = \begin{cases} 1 & \text{if } F(u,v) \leq F_0 \\ 0 & \text{if } F(u,v) > F_0 \end{cases}. \tag{4.8}$$

The image and graphical representation of an ideal low-pass filter are shown in Figure 4.7.

Because of the structure of the ILP mask, ringing occurs in the image when an ILP filter is applied to an image. ILP filter yields a blurred image, as shown in Figure 4.8.

4.2.1.2 Butterworth Low-Pass Filter (BLP)

This filter is used to eliminate high frequency noise with the least loss of image data in the specified pass band with order d. The transfer function of order d and with cutoff frequency F_0 can be expressed by using Equation 4.9:

$$P(u,v) = \frac{1}{1 + \left[F(u,v)/F_0\right]^{2d}}. \tag{4.9}$$

| Original Image | ILP Filter Radius: 5 | ILP Filter Radius: 15 | ILP Filter Radius: 30 | ILP Filter Radius: 80 |

FIGURE 4.8
ILP filter output with different values of F_0.

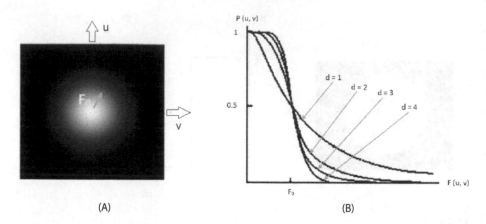

(A)   (B)

FIGURE 4.9
(A) Filter displayed as an image, (B) Graphical representation of BLP filter.

The image and graphical representation of a BLP filter are shown in Figure 4.9. The output of the BLP filter is shown in Figure 4.10.

4.2.1.3 Gaussian Low-Pass Filter (GLP)

The transfer function of a GLP filter is expressed in Equation 4.10:

$$P(u,v) = e^{-F^2(u,v)/2\sigma^2}. \tag{4.10}$$

Here, σ is the standard deviation and a measure of spread of the Gaussian curve. If σ is replaced with the cutoff radius F_0, then the transfer function of GLP is expressed as in Equation 4.11:

$$P(u,v) = e^{-F^2(u,v)/2F_0^2}. \tag{4.11}$$

The image and graphical representation of a GLP filter is shown in Figure 4.11.
The output of the GLP filter is shown in Figure 4.12.

Original Image   Order: 2,   Order: 2,   Order: 2,   Order: 2,
                Cutoff Radius: 10   Cutoff Radius: 30   Cutoff Radius: 70   Cutoff Radius: 150

FIGURE 4.10
Output of BLP filter with various cutoff radii.

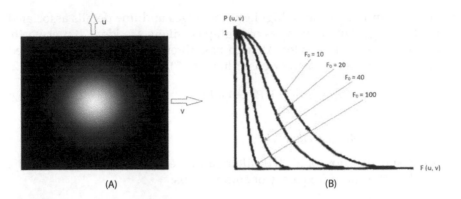

FIGURE 4.11
(A) Filter displayed as an image, (B) Graphical representation of GLP filter.

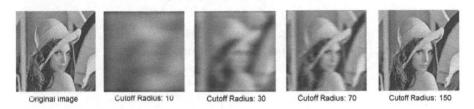

Original image   Cutoff Radius: 10   Cutoff Radius: 30   Cutoff Radius: 70   Cutoff Radius: 150

FIGURE 4.12
Output of GLP filter at different cutoff radius.

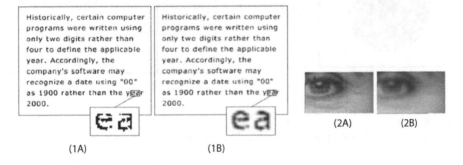

(1A)   (1B)

FIGURE 4.13
(1) Connecting text input-output: (1A) Input Image, (1B) Output of low-pass filter; (2) Blemishes reduction input–output: (2A) Input Image, (2B) Output of low-pass filter.

A low-pass filter can be used to connect broken text as well as reduce blemishes [18], as shown in Figure 4.13.

4.2.2 High Pass Filter

A high-pass filter suppresses frequencies lower than the cutoff frequency, but allows or passes high frequencies well [19]. A high-pass filter is generally used to sharpen an image and to highlight the edges and fine details associated with the image. Different types of high-pass filters are Ideal, Butterworth, and Gaussian high-pass filter. All high-pass filters (HPF) can be represented by their relationship to the low-pass filters (LPF), as shown in Equation 4.12:

$$HPF = 1 - LPF. \tag{4.12}$$

4.2.2.1 Ideal High-Pass Filter (IHP)

The transfer function of an IHP filter can be expressed by Equation 4.13, where F_0 is the cutoff frequency or cutoff radius:

$$P(u,v) = \begin{cases} 0 & \text{if } F(u,v) \leq F_0 \\ 1 & \text{if } F(u,v) > F_0 \end{cases}. \tag{4.13}$$

The image and graphical representation of an IHP filter is shown in Figure 4.14. The output of the IHP filter is shown in Figure 4.15.

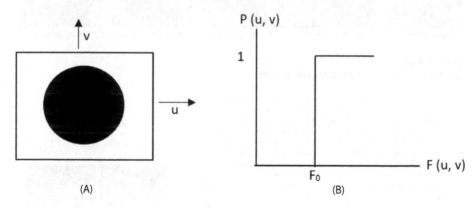

FIGURE 4.14
(A) Filter displayed as an image, (B) Graphical representation of IHP filter.

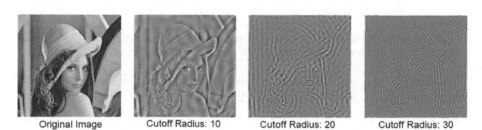

FIGURE 4.15
Output of IHP filter.

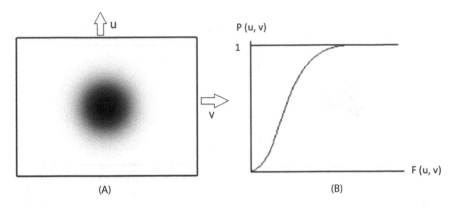

FIGURE 4.16
(A) Image representation of BHP, (B) Graphical representation of BHP.

4.2.2.2 Butterworth High-Pass Filter (BHP)

The transfer function of BHP filter can be defined by Equation 4.14 where p is the order and F_0 is the cutoff frequency or cutoff radius:

$$P(u,v) = \frac{1}{1 + \left[F_0 / F(u,v)\right]^{2n}}. \tag{4.14}$$

The image and graphical representation of BHP filter is shown in Figure 4.16. The output of the BHP filter is shown in Figure 4.17.

4.2.2.3 Gaussian High-Pass Filter (GHP)

The transfer function of GLP filter is expressed in Equation 4.15, with the cutoff radius F_0:

$$P(u,v) = 1 - e^{-F^2(u,v)/2F_0^2}. \tag{4.15}$$

The image and graphical representation of GLP filter is shown in Figure 4.18. The output of the GHP filter is shown in Figure 4.19.

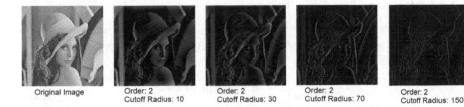

Original Image | Order: 2 Cutoff Radius: 10 | Order: 2 Cutoff Radius: 30 | Order: 2 Cutoff Radius: 70 | Order: 2 Cutoff Radius: 150

FIGURE 4.17
Output of BHP filter.

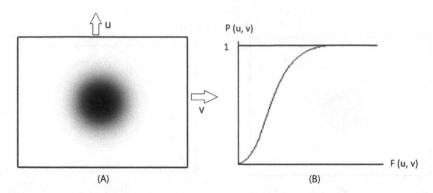

FIGURE 4.18
(A) Image representation of GHP, (B) Graphical representation of GHP.

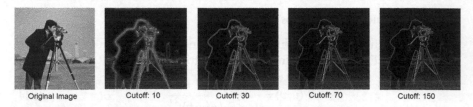

FIGURE 4.19
Output of GHP filter.

4.2.3 Band Pass Filter

A band pass suppresses very high and very low frequencies, but preserves an intermediate range band of frequencies [20]. Band pass filtering can be used to highlight edges (attenuating low frequencies) while decreasing the noise amount at the same time (suppressing high frequencies). To obtain the band pass filter function, the low-pass filter function is multiplied with the high-pass filter function in the frequency domain, where the cutoff frequency of the high pass is lower than that of the low pass. So, in theory, a band pass filter function can be developed if the low-pass filter function is available. The different types of band pass filter are Ideal band pass, Butterworth band pass, and Gaussian band pass.

4.2.3.1 Ideal Band Pass Filter (IBP)

The IBP allows the frequency within the pass band and removes the very high and very low frequency. An IBP filter within a frequency range $F_L, \ldots,$ F_H is defined by Equation 4.16:

$$P(u,v) = \begin{cases} 1 & \text{if } F_L \leq F(u,v) \leq F_H \\ 0 & \text{otherwise} \end{cases}. \tag{4.16}$$

Figure 4.20 shows the image and the effect of applying the IBP filter with different pass bands.

(A)   (B)   (C)   (D)   (E)

FIGURE 4.20
(A) Image of IBP filter, (B) Original image, (C) Output of IBP filter ($F_L = 30, F_H = 100$), (D) Output of IBP filter ($F_L = 30, F_H = 50$), and (E) Output of IBP filter ($F_L = 10, F_H = 90$).

4.2.3.2 Butterworth Band Pass Filter (BBP)

This filter can be obtained by multiplying the transfer function of a low and high Butterworth filter. If F_L is the low cutoff frequency, F_H is the high cutoff frequency, and p is the order of the filter then the BBP filter can be defined by Equation 4.17. The range of frequency is dependent on the order of the filter:

$$B_{LP}(u,v) = \frac{1}{1+\left[F(u,v)/F_L\right]^{2p}}$$

$$B_{HP}(u,v) = 1 - \frac{1}{1+\left[F(u,v)/F_H\right]^{2p}} \tag{4.17}$$

$$B_{BP}(u,v) = B_{LP}(u,v) * B_{HP}(u,v).$$

Figure 4.21 shows the image and the effect of applying the BBP filter with different pass bands and order $= 2$.

4.2.3.3 Gaussian Band Pass Filter (GBP)

This filter can be obtained by multiplying the transfer function of a low and high Gaussian filter. If F_L is the low cutoff frequency, F_H is the high cutoff frequency, and p is the order of the filter then the GBP filter can be defined by Equation 4.18:

(A)   (B)   (C)   (D)   (E)

FIGURE 4.21
(A) Image of BBP filter, (B) Original image, (C) Output of BBP filter ($F_L = 30, F_H = 50$), (D) Output of BBP filter ($F_L = 30, F_H = 150$), and (E) Output of BBP filter ($F_L = 70, F_H = 200$).

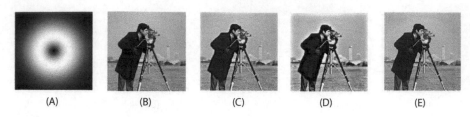

(A)   (B)   (C)   (D)   (E)

FIGURE 4.22
(A) Image of GBP filter, (B) Original image, (C) Output of GBP filter ($F_L = 30$, $F_H = 50$), (D) Output of GBP filter ($F_L = 10$, $F_H = 90$), and (E) Output of GBP filter ($F_L = 70$, $F_H = 90$).

$$G_{LP}(u,v) = e^{-F^2(u,v)/2F_0^2}$$
$$G_{HP}(u,v) = 1 - e^{-F^2(u,v)/2F_0^2} \tag{4.18}$$
$$G_{BP} = G_{LP}(u,v) * G_{HP}(u,v) \quad \text{where } F_L > F_H.$$

Figure 4.22 shows the image and the effect of applying the GBP filter with different pass bands.

4.2.4 Band Reject Filter

Band-reject filter (also called band-stop filter) is just the opposite of the bandpass filter [21]. It attenuates frequencies within a range of a higher and lower cutoff frequency. Different types of band reject filters are Ideal band reject, Butterworth band reject, and Gaussian band reject.

4.2.4.1 Ideal Band Reject Filter (IBR)

In this filter, the frequencies within the pass band are attenuated and the frequencies outside of the given range are passed without attenuation. Equation 4.19 defines an IBR filter with a frequency cutoff F_0, which is the center of the frequency band, and where W is the width of the frequency band:

$$P(u,v) = \begin{cases} 0 & \text{if } F_0 - \dfrac{W}{2} \leq F(u,v) \leq F_0 + \dfrac{W}{2} \\ 1 & \text{otherwise} \end{cases} \tag{4.19}$$

4.2.4.2 Butterworth Band Reject Filter (BBR)

In a BBR filter, frequencies at the center of the band are completely blocked. Frequencies at the edge of the frequency band are suppressed by a fraction of maximum value. If F_0 is the center of the frequency, W is the width of the frequency band, and p is the order of the filter, then a BBR filter can be defined by Equation 4.20:

$$P(u,v) = \frac{1}{1 + \left[F(u,v)W / (F(u,v)^2 - F_0^2) \right]^{2p}}. \tag{4.20}$$

4.2.4.3 Gaussian Band Reject Filter (GBR)

Here, the transition between the filtered and unfiltered frequency is very smooth. If F_0 is the center of the frequency and W is the width of the frequency band, then GBR filter can be defined by Equation 4.21:

$$P(u,v) = e^{-[(F(u,v)^2 - F_0^2 / F(u,v)W)]^2}. \tag{4.21}$$

4.3 Summary

Filtering is generally used to enhance the image detail. Several types of filters are discussed in this chapter. Mainly there are two types of filter: spatial and frequency. Spatial filtering is used to process the pixel value for the existing pixel, which is dependent on both itself and neighboring pixels. There are several types of spatial filter like linear, nonlinear, smoothing, and sharpening. Smoothing filter is used to blur the image and sharpening filter is used to highlight the blurred detail. Frequency filters are used to process the image in frequency domain. Different types of frequency filters are low-pass filters, which are used to blur the image, or high pass filters, which are used to highlight edges and sharpening. Band pass filtering can be used to highlight edges (attenuating low frequencies) and decrease the noise amount at the same time (suppressing high frequencies), while band reject filters are the opposite of band pass filters.

References

1. Araki, T., Ikeda, N., Dey, N., Acharjee, S., Molinari, F., Saba, L., Godia, E., Nicolaides, A., & Suri, J. S. 2015. Shape-based approach for coronary calcium lesion volume measurement on intravascular ultrasound imaging and its association with carotid intima-media thickness. *Journal of Ultrasound in Medicine*, 34(3), 469–482.

2. Gonzalez, R. C., Woods, R. E. 2016. Digital Image Processing 3rd edition, Prentice-Hall, New Jersey. ISBN-9789332570320, 9332570329.
3. Chaki, J., Parekh, R., & Bhattacharya, S. 2016. Plant leaf recognition using ridge filter and curvelet transform with neuro-fuzzy classifier. In *Proceedings of 3rd International Conference on Advanced Computing, Networking and Informatics* (pp. 37–44). Springer, New Delhi.
4. Santosh, K. C., Candemir, S., Jaeger, S., Karargyris, A., Antani, S., Thoma, G. R., & Folio, L. 2015. Automatically detecting rotation in chest radiographs using principal rib-orientation measure for quality control. *International Journal of Pattern Recognition and Artificial Intelligence*, 29(02), 1557001.
5. Ashour, A. S., Beagum, S., Dey, N., Ashour, A. S., Pistolla, D. S., Nguyen, G. N., et al. 2018. Light microscopy image de-noising using optimized LPA-ICI filter. *Neural Computing and Applications*, 29(12), 1517–1533.
6. Hangarge, M., Santosh, K. C., Doddamani, S., & Pardeshi, R. 2013. Statistical texture features based handwritten and printed text classification in south indian documents. *arXiv preprint arXiv:1303.3087*.
7. Dey, N., Ashour, A. S., Beagum, S., Pistola, D. S., Gospodinov, M., Gospodinova, E. P., & Tavares, J. M. R. 2015. Parameter optimization for local polynomial approximation based intersection confidence interval filter using genetic algorithm: An application for brain MRI image de-noising. *Journal of Imaging*, 1(1), 60–84.
8. Santosh, K. C., & Mukherjee, A. 2016, April. On the temporal dynamics of opinion spamming: Case studies on yelp. In *Proceedings of the 25th International Conference on World Wide Web* (pp. 369–379). International World Wide Web Conferences Steering Committee, Montréal, Québec, Canada.
9. Garg, A., & Khandelwal, V. 2018. Combination of spatial domain filters for speckle noise reduction in ultrasound medical images. *Advances in Electrical and Electronic Engineering*, 15(5), 857–865.
10. Nandi, D., Ashour, A. S., Samanta, S., Chakraborty, S., Salem, M. A., & Dey, N. 2015. Principal component analysis in medical image processing: A study. *International Journal of Image Mining*, 1(1), 65–86.
11. Kotyk, T., Ashour, A. S., Chakraborty, S., Dey, N., & Balas, V. E. 2015. Apoptosis analysis in classification paradigm: A neural network based approach. In *Healthy World Conference—A Healthy World for a Happy Life* (pp. 17–22). Kakinada (AP), India.
12. Santosh, K. C. 2010. Use of dynamic time warping for object shape classification through signature. *Kathmandu University Journal of Science, Engineering and Technology*, 6(1), 33–49.
13. Dhanachandra, N., Manglem, K., & Chanu, Y. J. 2015. Image segmentation using K-means clustering algorithm and subtractive clustering algorithm. *Procedia Computer Science*, 54(2015), 764–771.
14. Chakraborty, S., Chatterjee, S., Ashour, A. S., Mali, K., & Dey, N. 2018. Intelligent computing in medical imaging: A study. In *Advancements in Applied Metaheuristic Computing* (pp. 143–163). IGI Global, doi:10.4018/978-1-5225-4151-6.ch006.
15. Pardeshi, R., Chaudhuri, B. B., Hangarge, M., & Santosh, K. C. 2014, September. Automatic handwritten Indian scripts identification. In *14th International Conference on Frontiers in Handwriting Recognition (ICFHR), 2014* (pp. 375–380). IEEE.
16. Santosh, K. C., & Nattee, C. 2007. Template-based nepali natural handwritten alphanumeric character recognition. *Science & Technology Asia*, 12(1), 20–30.

17. Najarian, K., & Splinter, R. 2016. *Biomedical signal and image processing.* CRC Press, Boca Raton, FL.
18. Low-pass filter example [https://www.slideshare.net/SuhailaAfzana/image-smoothing-using-frequency-domain-filters (Last access date: June 10, 2018)]
19. Makandar, A., & Halalli, B. 2015. Image enhancement techniques using highpass and lowpass filters. *International Journal of Computer Applications,* 109(14).
20. Semmlow, J. L., & Griffel, B. 2014. *Biosignal and medical image processing.* CRC Press, Boca Raton, FL.
21. Konstantinides, K., & Rasure, J. R. 1994. The Khoros software development environment for image and signal processing. *IEEE Transactions on Image Processing,* 3(3), 243–252.

5

Segmentation Techniques

Image segmentation is the procedure of separating an image into several parts [1–3]. This is normally used to find objects or other significant information in digital images. There are various techniques to accomplish image segmentation discussed here.

5.1 Thresholding

Thresholding is a procedure of transforming an input grayscale image into a binarized image, or image with a new range of gray level, by using a particular threshold value [4,5]. The goal of thresholding is to extract some pixels from the image while removing others. The purpose of thresholding is to mark pixels that belong to foreground pixels with the same intensity and background pixels with different intensities.

Threshold is not only related to the image processing field. Rather threshold has the same meaning in any arena. A threshold is basically a value having two set of regions on its either side, that is, above the threshold or below the threshold. Any function can have a threshold value [6]. The function has different expressions for below the threshold value and for above the threshold value. For an image, if the pixel value of the original image is less than or below a particular threshold value it will follow a specific transformation or conversion function, if not, it will follow another. Threshold can be global or local. Global threshold means the threshold is selected from the whole image. Local or adaptive threshold is used when the image has uneven illumination, which makes it difficult to segment using a single threshold. In that case, the original image is divided into subimages, and for each subimage a particular threshold is used for segmentation [7]. Figure 5.1 shows the segmentation output with local and global threshold.

5.1.1 Histogram Shape-Based Thresholding

The histogram method presumes that there is some average value for the foreground or object pixels and background, but the reality is that the real pixel values have some deviation around these average values [8,9]. In that case, selecting an accurate image threshold value is difficult and computationally

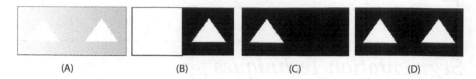

(A)   (B)   (C)   (D)

FIGURE 5.1
(A) Input image with uneven illumination, (B) and (C) Global thresholding result, (D) Local thresholding result.

expensive. One comparatively simple technique is the iterative method to find a specific image threshold, which is also robust against noise. The steps of the iterative method is as follows:

Step 1: An initial threshold (T) is selected arbitrarily by any other desired method.

Step 2: The image $I(x, y)$ is segmented into foreground or object pixels and background pixels:

$$\text{Object pixels (OP)} \leftarrow \{I(x,y) : I(x,y) \geq T\}$$
$$\text{Background pixels (BP)} \leftarrow \{I(x,y) : I(x,y) < T\}. \tag{5.1}$$

Step 3: The average of each pixel set is calculated.

$$A_{OP} \leftarrow \text{Average of OP}$$

$$A_{BP} \leftarrow \text{Average of BP}$$

Step 4: A new threshold is formed, which is the average of A_{OP} and A_{BP}:

$$T_{new} \leftarrow \frac{(A_{OP} + A_{BP})}{2}. \tag{5.2}$$

Step 5: In step 2, use the new threshold obtained in step 4. Repeat till the new threshold matches the one before it.

Assume that the gray level image $I(x, y)$ is composed of a light object in a dark background, in such a way that background and object, or foreground gray level pixels, can be grouped into two dominant modes. One clear way to extract the object pixels from the background is to select a threshold T, which divides these two modes. Then any pixel (x, y) where $I(x, y) \geq T$ is called an object pixel, otherwise, the pixel is called a background pixel. Example:

If two dominant modes describe the image histogram, it is called a bimodal histogram. Here, only one threshold is sufficient for segmenting or partitioning the image. Figure 5.2 shows the bimodal histogram of an image and the segmented image.

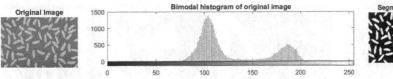

FIGURE 5.2
The bimodal histogram and the segmented image.

If for instance, an image is composed of two or more types of dark objects in a light background, three or more dominant modes are used to characterize the image histogram, which is denoted as a multimodal histogram. Figure 5.3 shows the multimodal histogram of an image and the segmented image.

5.1.2 Clustering-Based Thresholding

K-means Thresholding Method: The steps of *K*-means algorithm for selecting the threshold is as follows [10]:

Step 1: Class centers (*K*) are initialized:

$$C_{j0} = G_{\min} + \left[\frac{(j-(j/2))(G_{\max} - G_{\min})}{k} \right],$$ (5.3)

where $j = 1,2,\dots,k$; C_{j0} is the first class center of *j*th class; $G_{\min}$ and $G_{\max}$ are the minimum and maximum gray value of the sample space.

Step 2: Assign every point of the sample space to its nearest class center based on Euclidean Distance:

$$D_{j,i} = \text{abs}(G_i - C_j),$$ (5.4)

where $j = 1,2,\dots,k$; $i = 1,2,\dots,P$; $D_{j,i}$ is the distance from an *i*th point to the *j*th class, and *P* is the total number of points in the sample space.

Step 3: Compute the (*K*) new class centers from the average of the points that are assigned to it:

$$C_{\text{new}} = \frac{1}{Pi} \sum G_j,$$ (5.5)

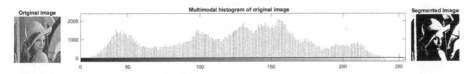

FIGURE 5.3
The multimodal histogram and the segmented image.

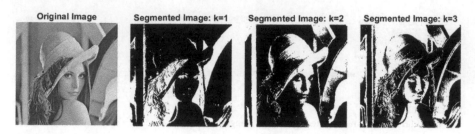

FIGURE 5.4
The output of image segmentation with different *k* values.

where $j = 1,2,...,k$ and Pi is the total number of points that are assigned to the *i*th class in step 2.

Step 4: Repeat step 2 for change in the class center; otherwise stop the iteration.

Step 5: The threshold is calculated by the mean of the *k*th class center and $(k-1)$ class center:

$$T = \frac{1}{2}(C_k + C_{k-1}). \tag{5.6}$$

The result of the image segmentation is shown in Figure 5.4.

Otsu-Clustering Thresholding Method: This method is used to select a threshold value by minimizing the within class variances of two clusters [11,12]. The within-class variance can be expressed by Equation 5.7:

$$\sigma_w^2(T) = P_b(T)\sigma_b^2(T) + P_f(T)\sigma_f^2(T), \tag{5.7}$$

where P_f and P_b are the probability of foreground and background class occurrences; T is the initial threshold value, which is randomly selected by some algorithm, and σ_f^2 and σ_b^2 are the variances of foreground and background clusters.

The probability of foreground and background class occurrences can be denoted by Equation 5.8:

$$P_b(T) = \sum_{G=0}^{T} p(G)$$

$$P_f(T) = \sum_{G=T+1}^{L-1} p(G), \tag{5.8}$$

where G is the gray level values $\{0,1,...,L-1\}$ and $p(G)$ is the probability mass function of G.

The variances of foreground and background clusters are defined by Equation 5.9:

$$\sigma_b^2(T) = \sum_{G=0}^{T}(G - M_b(T))^2 \frac{p(G)}{P_b(T)}$$

$$\sigma_f^2(T) = \sum_{G=T+1}^{L-1}(G - M_f(T))^2 \frac{p(G)}{P_f(T)},$$

(5.9)

where M_b and M_f are the means of background and foreground clusters respectively and can be defined by Equation 5.10:

$$M_f(T) = \sum_{G=0}^{T}G \times p(G)$$

$$M_b(T) = \sum_{G=T+1}^{L-1}G \times p(G).$$

(5.10)

A lot of computations are involved in computing the within class variance for each of the two classes for every possible threshold. Thus, the between-class variance is computed by subtracting the within class variance from the total variance:

$$\sigma_{between}^2(T) = \sigma_{total}^2(T) - \sigma_w^2(T)$$

$$= P_b(T)[\mu_b(T) - M_{total}]^2 + P_f(T)[\mu_f(T) - M_{total}]^2.$$

(5.11)

σ_{total}^2 and M_{total} can be expressed by the Equation 5.12:

$$\sigma_{total}^2 = \sum_{G=0}^{L-1}(G - M_{total})^2 p(G)$$

$$M_{total} = \sum_{G=0}^{L-1}G \times p(G).$$

(5.12)

The main advantage of this method is its simple computation.

Figure 5.5 shows the segmented output using a different number of clusters.

FIGURE 5.5
Segmented output using the Otsu clustering thresholding method.

5.1.3 Entropy-Based Thresholding

This method is created based on the probability distribution function of the gray level histogram [13,14]. Two entropies can be calculated: one for black pixels and the other for white pixels:

$$\sum_{i=0}^{255} g(i) = 1$$

$$E_b(t) = -\sum_{i=0}^{t} \frac{g(i)}{\sum_{j=0}^{t} g(j)} * \log \frac{g(i)}{\sum_{j=0}^{t} g(j)} \tag{5.13}$$

$$E_w(t) = -\sum_{i=t+1}^{255} \frac{g(i)}{\sum_{j=t+1}^{255} g(j)} * \log \frac{g(i)}{\sum_{j=t+1}^{255} g(j)},$$

where $g(i)$ is the normalized histogram.

The optimal single threshold value is selected by maximizing the entropy of black and white pixels, and can be depicted by Equation 5.14:

$$T = \text{Arg} \max_{t=0\ldots255} E_b(t) + E_w(t). \tag{5.14}$$

p optimal threshold values can be found by Equation 5.15:

$$\{T_1,\ldots,T_p\} = \text{Arg} \max_{t_1 < \cdots < t_p} E(-1,t_1) + E(t_1,t_2) + \cdots + E(t_p, 255), \tag{5.15}$$

where

$$E(t_n, t_{n+1}) = -\sum_{i=t_n+1}^{t_{n+1}} \frac{g(i)}{\sum_{j=t_n+1}^{t_{n+1}} g(j)} \log \frac{g(i)}{\sum_{j=t_n+1}^{t_{n+1}} g(j)}.$$

Figure 5.6 shows the output of the segmented image using an entropy-based method.

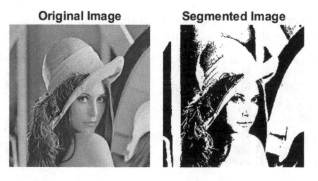

Original Image Segmented Image

FIGURE 5.6
Segmented output using entropy-based method.

5.2 Edge-Based Segmentation

Edge segmentation is a vital area of research, as it helps higher-level image exploration [15]. Detection of edges is an important tool for image segmentation. The representation of the edge of an image meaningfully decreases the amount of data to be processed, however, it holds vital information about the shapes of objects in the scene [16]. Edges are basically local variations in image intensity. Edge detection approaches convert original images into edge images depending on the variations of gray tones in the image. Image edge detection is used in many applications like object shape identification, medical image processing, biometrics, and so on [17,18]. There are three different types of discontinuities in the gray level such as points, lines, and edges. Spatial masks can be used to identify these three types of image discontinuities.

5.2.1 Roberts Edge Detector

The Roberts edge detection technique is used to highlight high spatial frequency regions of the image, which corresponds to edges. The input to the operator is a grayscale image [19]. A mask is used to compute the output, which operates on each pixel values of the input image. Figure 5.7 shows the values of the mask. Here, M_x is the mask used in the horizontal direction and M_y is the mask used in the vertical direction.

Figure 5.8 shows the detected edge output using a Roberts edge detector.

-1	0		0	-1
0	+1		+1	0

$$M_x \qquad\qquad M_y$$

FIGURE 5.7
Masks used in Roberts edge detection.

Original Image   Roberts Edge Detection

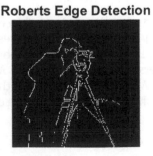

FIGURE 5.8
Edge detection using the Roberts edge detector.

-1	-2	-1
0	0	0
+1	+2	+1

$$M_x$$

-1	0	+1
-2	0	+2
-1	0	+1

$$M_y$$

FIGURE 5.9
Masks used in Sobel edge detection.

FIGURE 5.10
Edge detection using the Sobel edge detector.

5.2.2 Sobel Edge Detector

The Sobel edge detection method uses the Sobel approximation to the derivative to highlight edges [19,20]. It leads the edges at those points where the gradient is highest. The operator comprises of a pair of 3×3 kernels, or masks, as shown in Figure 5.9. One kernel is simply the 90° rotated version of other. Here, M_x is the mask used in the horizontal direction and M_y is the mask used in the vertical direction.

Figure 5.10 shows the detected edge output using a Sobel edge detector.

5.2.3 Prewitt Edge Detector

The Prewitt edge detection is used to assess the orientation and magnitude of an edge [19]. The operator comprises of a pair of 3×3 kernels, or masks, as shown in Figure 5.11. Like the Sobel operator, one kernel is simply the 90°-rotated version of the other. Here, M_x is the mask used in the horizontal direction and M_y is the mask used in the vertical direction.

Figure 5.12 shows the detected edge output using a Prewitt edge detector.

5.2.4 Kirsch Edge Detector

Kirsch edge detection uses a single mask and rotates it to eight directions: North, West, East, South, Northwest, Southwest, Southeast, and Northeast [21].

-1	-1	-1
0	0	0
+1	+1	+1

-1	0	+1
-1	0	+1
-1	0	+1

$$M_x \qquad\qquad M_y$$

FIGURE 5.11
Masks used in Prewitt edge detection.

Original Image

Prewitt Edge Detection

FIGURE 5.12
Edge detection using the Prewitt edge detector.

$$N = \begin{bmatrix} 5 & 5 & 5 \\ -3 & 0 & -3 \\ -3 & -3 & -3 \end{bmatrix} \quad W = \begin{bmatrix} 5 & -3 & -3 \\ 5 & 0 & -3 \\ 5 & -3 & -3 \end{bmatrix} \quad E = \begin{bmatrix} -3 & -3 & 5 \\ -3 & 0 & 5 \\ -3 & -3 & 5 \end{bmatrix} \quad S = \begin{bmatrix} -3 & -3 & -3 \\ -3 & 0 & -3 \\ 5 & 5 & 5 \end{bmatrix}$$

$$NW = \begin{bmatrix} 5 & 5 & -3 \\ 5 & 0 & -3 \\ -3 & -3 & -3 \end{bmatrix} \quad NE = \begin{bmatrix} -3 & 5 & 5 \\ -3 & 0 & 5 \\ -3 & -3 & -3 \end{bmatrix} \quad SE = \begin{bmatrix} -3 & -3 & 5 \\ -3 & 0 & 5 \\ -3 & 5 & 5 \end{bmatrix} \quad SW = \begin{bmatrix} -3 & -3 & -3 \\ 5 & 0 & -3 \\ 5 & 5 & -3 \end{bmatrix}$$

FIGURE 5.13
Masks used in Kirsch edge detection.

The edge magnitude is denoted as the maximum value found by convolution of each mask with the image. The masks are defined as shown in Figure 5.13.

Figure 5.14 shows the detected edge output using a Kirsch edge detector.

5.2.5 Robinson Edge Detector

The Robinson method is implemented by using coefficients of 0, 1, and 2 [22]. The masks are symmetrical around their directional axis, which is composed of zeros. The edge magnitude is the maximum value obtained by convolving the mask with the image pixel neighborhood, and the edge angle can be obtained by the angle of the line of zeroes in the mask containing the maximum response. The masks are shown in Figure 5.15.

Figure 5.16 shows the detected edge output using a Robinson edge detector.

Original Image   **Kirsch Edge Detection**

FIGURE 5.14
Edge detection using the Kirsch edge detector.

$$N = \begin{bmatrix} 1 & 2 & 1 \\ 0 & 0 & 0 \\ -1 & -2 & -1 \end{bmatrix} \quad W = \begin{bmatrix} 1 & 0 & -1 \\ 2 & 0 & -2 \\ 1 & 0 & -1 \end{bmatrix} \quad E = \begin{bmatrix} -1 & 0 & 1 \\ -2 & 0 & 2 \\ -1 & 0 & 1 \end{bmatrix} \quad S = \begin{bmatrix} -1 & -2 & -1 \\ 0 & 0 & 0 \\ 1 & 2 & 1 \end{bmatrix}$$

$$NW = \begin{bmatrix} 2 & 1 & 0 \\ 1 & 0 & -1 \\ 0 & -1 & -2 \end{bmatrix} \quad NE = \begin{bmatrix} 0 & 1 & 2 \\ -1 & 0 & 1 \\ -2 & -1 & 0 \end{bmatrix} \quad SE = \begin{bmatrix} -2 & -1 & 0 \\ -1 & 0 & 1 \\ 0 & 1 & 2 \end{bmatrix} \quad SW = \begin{bmatrix} 0 & -1 & -2 \\ 1 & 0 & -1 \\ 2 & 1 & 0 \end{bmatrix}$$

FIGURE 5.15
Masks used in Robinson edge detection.

Original Image   **Robinson Edge Detection**

FIGURE 5.16
Edge detection using the Robinson edge detector.

5.2.6 Canny Edge Detector

A canny edge detector is used to find the edge of an image by separating noise from the image prior to edge extraction [23,24]. The steps are as follows:

Step 1: The image $I(x, y)$ is convolved with a Gaussian function G to reduce the noise and to get a smooth version of the image:

$$S(x,y) \leftarrow I(x,y) * G. \tag{5.16}$$

Original Image **Canny Edge Detection**

FIGURE 5.17
Edge detection using a Canny edge detector.

Step 2: Gradient magnitude and direction is calculated for every pixel of $S(x, y)$, as obtained before.

Step 3: Nonmaximal suppression is applied to the gradient magnitude.

Step 4: Finally, a threshold is applied to the nonmaximal suppression image.

Figure 5.17 shows the detected edge output using Canny edge detector.

5.2.7 Laplacian of Gaussian (LoG) Edge Detector

LoG of an image $I(x, y)$ is defined [25] by a second order derivative defined as:

$$\nabla^2 I = \frac{\partial^2 I}{\partial x^2} + \frac{\partial^2 I}{\partial y^2}. \tag{5.17}$$

It smoothes the image and the Laplacian is also calculated. This results in a double edge image. Zero crossings are found out from the filtered image to find the edge. This operator is used to find a pixel in the light or dark side of the edge. The masks used in this operation are shown in Figure 5.18 where M_x and M_y are the masks used in the horizontal and vertical direction.

Figure 5.19 shows the detected edge output using an LoG edge detector.

0	-1	0
-1	4	-1
0	-1	0

M_x

-1	-1	-1
-1	8	-1
-1	-1	-1

M_y

FIGURE 5.18
Masks used in LoG edge detection.

Original Image   **LoG Edge Detection**

FIGURE 5.19
Edge detection using the LoG edge detector.

5.2.8 Marr-Hildreth Edge Detection

The Marr-Hildreth method is a technique of highlighting edges in continuous curves wherever there are quick variations in image brightness [26]. The LoG function is used as the convolution function. Then zero-crossings are found in the filtered result to find the edges. Algorithmic steps for the Marr-Hildreth edge detector are as follows:

Step 1: The image is convolved with the Gaussian function for smoothing

Step 2: 2D Laplacian is applied to the smoothed image

Step 3: Analyze the sign change by looping through the result

Step 4: If a sign change occurs, and if the slope across the sign change is greater than a threshold, then consider it as an edge.

Figure 5.20 shows the detected edge output using a Marr-Hildreth edge detector.

Original image   **Marr-Hildreth Edge Detection**

FIGURE 5.20
Edge detection using the Marr-Hildreth edge detector.

5.3 Region-Based Segmentation

Region-based segmentation is used to split or merge regions in the image based on some similar or common image properties such as intensity values of the region, texture, or pattern of the region and so on [27,28]. This can be divided into two main types: region growing or region merging and region splitting.

5.3.1 Region Growing or Region Merging

This is a method to group, or merge, pixels or subregions into larger regions based on some property or image attributes [29]. This method starts with a set of seed points and is based on some similar properties such as texture, gray level, shape, color, and so forth; the neighboring pixels are appended or added with the seed region. One such method is to divide the image into 2×2 or 4×4 regions and check each one. In the worst case the seed can be a single pixel. Merging is done until no neighboring pixels are left with the same property. Finally, the region is extracted from the image and a new seed is defined to merge other similar regions. If the homogeneous regions are small, region growing or merging is preferred.

5.3.2 Region Splitting

This method is just the opposite of region growing [30]. Region splitting starts with the whole image and is divided until a uniform subregion is found. The main drawback of this method is that it is very difficult to find a proper location to make the partition. If the homogeneous regions are large, region splitting is preferred.

5.4 Summary

Image segmentation is generally used to separate an image into different parts, or to extract significant information from the image. Thresholding is one of the procedures for image segmentation. The goal of thresholding is to extract some pixels from the image while removing others. There are different methods to select the threshold value like a histogram shape-based method, clustering-based method, and entropy-based method. In the histogram shape-based method peaks, valleys, and curvatures of the histogram are analyzed. In a clustering-based method, the image is clustered into different parts based on the pixel values. There are different clustering-based methods like k-means, or Otsu. The entropy-based method is developed based on the probability distribution function of the gray level histogram. There are different edge-based segmentations like Sobel, Canny, Prewitt, Robinson, Robert, Kirsch, LoG, and Marr-Hildreth. The Roberts edge detection technique is used to highlight high spatial frequency regions of the image, which corresponds to edges. The Sobel edge detection method uses the Sobel approximation to the derivative to highlight edges. The Prewitt edge detection is used to assess the orientation and magnitude of an edge. The Kirsch edge detection use a single mask and rotates it to eight directions: North, West, East, South, Northwest, Southwest, Southeast, and Northeast. The Robinson method is same as the Kirsch method but the only difference is that it is implemented by using the coefficients of 0, 1, and 2. A Canny edge detector is used to find the edge of the image by separating noise from the image prior to edge extraction. The LoG operator is used to find a pixel in the light or the dark side of the edge. The Marr-Hildreth method is a technique of highlighting edges in continuous curves wherever there are fast variations in image brightness. Segmentation can be done based on region properties. In region growing, or merging, first a small region is taken, then based on some similar property the neighboring pixels are added to that region. In region splitting, the whole image is split into sun regions that satisfy the predefined similarity or homogeneity property.

References

1. Santosh, K. C., Xue, Z., Antani, S. K., & Thoma, G. R. 2015. NLM at imageCLEF2015: Biomedical multipanel figure separation. In *CLEF (Working Notes)*, 1391, 1–8. ISSN 1613-0073.

2. Roy, P., Goswami, S., Chakraborty, S., Azar, A. T., & Dey, N. 2014. Image segmentation using rough set theory: A review. *International Journal of Rough Sets and Data Analysis (IJRSDA)*, 1(2), 62–74.

3. Obaidullah, S. M., Halder, C., Santosh, K. C., Das, N., & Roy, K. 2018. PHDIndic_11: Page-level handwritten document image dataset of 11 official Indic scripts for script identification. *Multimedia Tools and Applications*, 77(2), 1643–1678.

4. Araki, T., Ikeda, N., Dey, N., Acharjee, S., Molinari, F., Saba, L., Godia, E., Nicolaides, A., & Suri, J. S. 2015. Shape-based approach for coronary calcium lesion volume measurement on intravascular ultrasound imaging and its association with carotid intima-media thickness. *Journal of Ultrasound in Medicine*, 34(3), 469–482.

5. Chaki, J., & Parekh, R. 2011. Plant leaf recognition using shape based features and neural network classifiers. *International Journal of Advanced Computer Science and Applications*, 2(10), 41–47.

6. Chaki, J., Parekh, R., & Bhattacharya, S. 2015, July. Recognition of whole and deformed plant leaves using statistical shape features and neuro-fuzzy classifier. In *Recent Trends in Information Systems (ReTIS), 2015 IEEE 2nd International Conference on* (pp. 189–194). IEEE, Kolkata, India.

7. Santosh, K. C., & Antani, S. 2018. Automated chest x-ray screening: Can lung region symmetry help detect pulmonary abnormalities? *IEEE Transactions on Medical Imaging*, 37(5), 1168–1177.

8. Roy, P., Dutta, S., Dey, N., Dey, G., Chakraborty, S., & Ray, R. 2014, July. Adaptive thresholding: A comparative study. In *Control, Instrumentation, Communication and Computational Technologies (ICCICCT), 2014 International Conference on* (pp. 1182–1186). IEEE.

9. Chaki, J., Parekh, R., & Bhattacharya, S. 2015. Plant leaf recognition using texture and shape features with neural classifiers. *Pattern Recognition Letters*, 58, 61–68.

10. Dhanachandra, N., Manglem, K., & Chanu, Y. J. 2015. Image segmentation using K-means clustering algorithm and subtractive clustering algorithm. *Procedia Computer Science*, 54(2015), 764–771.

11. Satapathy, S. C., Raja, N. S. M., Rajinikanth, V., Ashour, A. S., & Dey, N. 2016. Multi-level image thresholding using Otsu and chaotic bat algorithm. *Neural Computing and Applications*, 29(12), 1–23.

12. Chaki, J., Parekh, R., & Bhattacharya, S. 2016, January. Plant leaf recognition using a layered approach. In *Microelectronics, Computing and Communications (MicroCom), 2016 International Conference on* (pp. 1–6). IEEE, Durgapur, India.

13. Shriranjani, D., Tebby, S. G., Satapathy, S. C., Dey, N., & Rajinikanth, V. 2018. Kapur's entropy and active contour-based segmentation and analysis of retinal optic disc. In *Computational Signal Processing and Analysis* (pp. 287–295). Springer, Singapore.

14. Chakraborty, S., Chatterjee, S., Dey, N., Ashour, A. S., Ashour, A. S., Shi, F., & Mali, K. 2017. Modified cuckoo search algorithm in microscopic image segmentation of hippocampus. *Microscopy Research and Technique*, 80(10), 1051–1072.

15. Chaki, J., & Parekh, R. 2012. Designing an automated system for plant leaf recognition. *International Journal of Advances in Engineering & Technology*, 2(1), 149.

16. Dey, N., Pal, M., & Das, A. 2012. A session based blind watermarking technique within the NROI of retinal fundus images for authentication using DWT, spread spectrum and harris corner detection. *arXiv preprint arXiv:1209.0053*.

17. Santosh, K. C., & Nattee, C. 2007. Template-based nepali natural handwritten alphanumeric character recognition. *Science & Technology Asia*, 12(1), 20–30.

18. Samanta, S., Acharjee, S., Mukherjee, A., Das, D., & Dey, N. 2013, December. Ant weight lifting algorithm for image segmentation. In *Computational Intelligence and Computing Research (ICCIC), 2013 IEEE International Conference on* (pp. 1–5). IEEE.

19. Ganesan, P., & Sajiv, G. 2017, March. A comprehensive study of edge detection for image processing applications. In *Innovations in Information, Embedded and Communication Systems (ICIIECS), 2017 International Conference on* (pp. 1–6). IEEE.

20. Biswas, D., Das, P., Maji, P., Dey, N., & Chaudhuri, S. S. 2013. Visible watermarking within the region of noninterest of medical images based on fuzzy C-means and Harris corner detection. *Computer Science & Information Technology*, 161–168.

21. Melin, P., Gonzalez, C. I., Castro, J. R., Mendoza, O., & Castillo, O. 2014. Edge-detection method for image processing based on generalized type-2 fuzzy logic. *IEEE Transactions on Fuzzy Systems*, 22(6), 1515–1525.

22. Russ, J. C. 2016. *The Image Processing Handbook*. CRC Press, Boca Raton, FL.

23. Chaki, J., Parekh, R., & Bhattacharya, S. 2016. Plant leaf recognition using ridge filter and curvelet transform with neuro-fuzzy classifier. In *Proceedings of 3rd International Conference on Advanced Computing, Networking and Informatics* (pp. 37–44). Springer, New Delhi.

24. Dey, N., Maji, P., Das, P., Biswas, S., Das, A., & Chaudhuri, S. S. 2013, January. An edge based blind watermarking technique of medical images without devalorizing diagnostic parameters. In *Advances in Technology and Engineering (ICATE), 2013 International Conference on* (pp. 1–5). IEEE.
25. Thamotharan, B., Venkatraman, B., Anusuya, A., Ramakrishnan, S., & Karthikeyan, M. P. 2017. Analysis of various edge detection techniques for dosimeter bubble detector images. *Biomedical Research*, 28(20), 8635–8639.
26. Hemalatha, R., Santhiyakumari, N., Madheswaran, M., & Suresh, S. 2017, March. Intima-media segmentation using marr-hildreth method and its implementation on unified technology learning platform. In *Emerging Devices and Smart Systems (ICEDSS), 2017 Conference on* (pp. 32–36). IEEE.
27. Chakraborty, S., Chatterjee, S., Ashour, A. S., Mali, K., & Dey, N. 2018. Intelligent computing in medical imaging: A study. In *Advancements in Applied Metaheuristic Computing* (pp. 143–163). IGI Global.
28. Dey, N., Rajinikanth, V., Ashour, A. S., & Tavares, J. M. R. 2018. Social group optimization supported segmentation and evaluation of skin melanoma images. *Symmetry*, 10(2), 51.
29. Hore, S., Chakraborty, S., Chatterjee, S., Dey, N., Ashour, A. S., Van Chung, L., & Le, D. N. 2016. An integrated interactive technique for image segmentation using stack based seeded region growing and thresholding. *International Journal of Electrical and Computer Engineering*, 6(6), 2773.
30. Ohlander, R., Price, K., & Reddy, D. R. 1978. Picture segmentation using a recursive region splitting method. *Computer Graphics and Image Processing*, 8(3), 313–333.

6

Mathematical Morphology Techniques

6.1 Binary Morphology

Binary images generally contain several flaws [1]. The binary regions which are created by simple thresholding can contain noise. Morphological image processing is used to remove these imperfections of the image. Morphological techniques use a small shape or template called a structuring element. The structuring element is placed in all possible positions in the image and is compared with the corresponding neighborhood of pixels. Some procedures check whether the structuring element "fits" inside the neighborhood, while others check whether it intersects or "hits" the neighborhood. Figure 6.1 shows the effect of a structuring element on image pixel.

Through the binary morphological operation on a binary image [2], another new binary image is created which contains nonzero pixel values only if the test is successful for that particular location of the input image. The structuring element is a small matrix of pixels, that is, a small binary image, with values of either zero or one. The arrangement of zeroes and ones denotes the shape of the structuring element. The origin of the structuring element is generally one of its pixels. The size of the structuring element is usually odd in dimension, and the center pixel is considered as the origin of the structuring element. Figure 6.2 shows some examples of structuring elements.

When a structuring element is positioned onto a binary image, each pixel of the binary image is associated with the pixels of the structuring element. The structuring element is considered to fit the image if, for all of its pixels containing the value 1, the associated image pixel is also 1. Likewise, a structuring element is considered to intersect or hit an image if at least for one of its pixels containing the value 1 the associated image pixel is also 1. Figure 6.3 demonstrates the effect of hit and fit.

The structuring elements having zero valued pixels are ignored.

6.1.1 Erosion

The erosion operation [3,4] (denoted by $\ominus$) between a binary image $I(x, y)$ and a structuring element S, creates a new binary image $E(x, y)$ with ones in each

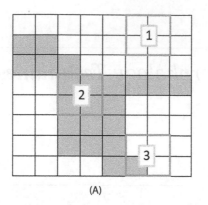

(A)

(B)

FIGURE 6.1
(A) White pixels contain zero value, and nonzero pixels contain nonzero value, (1) The structuring element neither fits, nor hits the image, (2) The structuring element fits the image, (3) The structuring element hits the image; (B) structuring element.

1	1	1	1	1
1	1	1	1	1
1	1	1	1	1
1	1	1	1	1
1	1	1	1	1

(A)

0	0	1	0	0
0	1	1	1	0
1	1	1	1	1
0	1	1	1	0
0	0	1	0	0

(B)

0	0	1	0	0
0	0	1	0	0
1	1	1	1	1
0	0	1	0	0
0	0	1	0	0

(C)

1	1	1
1	1	1
1	1	1

(D)

FIGURE 6.2
Orange pixel is the origin of the structuring element, (A) 5 × 5 square-shaped structuring element, (B) 5 × 5 diamond-shaped structuring element, (C) 5 × 5 cross-shaped structuring element, (D) 3 × 3 square-shaped structuring element.

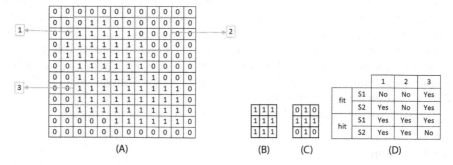

FIGURE 6.3
(A) Superimposing of structuring element, (B) Structuring element S1, (C) Structuring element S2, and (D) Fitting and hitting of a binary image with structuring elements S1 and S2.

FIGURE 6.4
Output of erosion of a binary image.

and every locations (x, y) at which that structuring element S fits the input image I and zero otherwise:

$$E(x,y) = I \ominus S. \tag{6.1}$$

Erosion with a small (e.g., 3×3 or 5×5) structuring element shrinks an image by removing a pixel layer from both the outer and inner boundaries of image regions. So, image details are removed and different gaps and holes of the image regions become larger. Figure 6.4 shows the output of erosion with a 5×5 structuring element.

Erosion with a large structuring element has a more prominent effect. The erosion result with a large structuring element can be the same if a smaller structuring element of the same shape is iteratively applied to an image. For example, if $S1$ and $S2$ are two structuring elements duplicate in shape, but the size of $S2$ is twice that of $S1$, then

$$I \ominus S2 = (I \ominus S1) \ominus S1. \tag{6.2}$$

By using erosion small details of the image can be removed, and thus the size of the region of interest can be reduced. The boundary (B) of an image region can be obtained by subtracting the eroded image from the original image:

$$B = I - (I \ominus S). \tag{6.3}$$

6.1.2 Dilation

The dilation operation [5,6] (denoted by $\oplus$) between a binary image $I(x, y)$ and a structuring element S, creates a new binary image $D(x, y)$ with ones in each and every location (x, y) where that structuring element S hits the input image I and 0:

$$E(x,y) = I \oplus S. \tag{6.4}$$

Thus, dilation is just the opposite of erosion. It adds a pixel layer to both the outer and inner boundaries of image regions. Figure 6.5 shows the output of dilation with a 5×5 structuring element.

FIGURE 6.5
Output of dilation of a binary image.

By using dilation, the holes of a binary image can be filled and gaps between different regions can be reduced.

6.1.3 Opening

The opening operation [7] (denoted by ∘) between a binary image $I(x, y)$ and a structuring element S, creates a new binary image $O(x, y)$, which is basically erosion followed by dilation:

$$I \circ S = (I \ominus S) \oplus S. \tag{6.5}$$

The opening is used to open up gaps between connected regions within the image. Figure 6.6 shows the output of opening with 7×7 structuring element.

Once the connected regions within the image are opened by using a structuring element, there is no further opening effect on that image using that particular structuring element:

$$I \circ S = (I \circ S) \circ S. \tag{6.6}$$

6.1.4 Closing

The closing operation [7] (denoted by ·) between a binary image $I(x, y)$ and a structuring element S, creates a new binary image $C(x, y)$, which is basically dilation followed by erosion:

$$I \cdot S = (I \oplus S) \ominus S. \tag{6.7}$$

A closing operation is used to connect or fill holes in the image regions while maintaining the initial region sizes. Figure 6.7 shows the output of closing with a 7×7 structuring element.

FIGURE 6.6
Output of opening of a binary image.

Original Image   **Binary Image**   Closing Output

FIGURE 6.7
Output of closing of a binary image.

Once the holes are connected within the image by using a structuring element, there is no further closing effect on that image using that particular structuring element:

$$I \cdot S = (I \cdot S) \cdot S. \tag{6.8}$$

6.1.5 Hit and Miss

The hit and miss operation [8] (denoted as $\odot$) permits to develop information on how objects in a binary image are associated to their neighbors. The operation calls for a matched pair of structuring elements, $\{S1, S2\}$, that investigate the outside and inside, individually, of objects in the image:

$$I \odot \{S1, S2\} = (I \ominus S1) \cap (I^C \ominus S2). \tag{6.9}$$

Here, I^C is the complement of I.

An object pixel is conserved by this operation if and most effective if $S1$ transformed to that pixel fits inside the object and $S2$ transformed to that pixel fits outside the object. Figure 6.8 shows the output of a hit and miss operation. It is assumed that $S1 \cap S2 = NULL$. This operation is generally used for detecting particular shapes where two structuring elements present that particular shape.

6.1.6 Thinning

The thinning operation is to some extent like opening or erosion [8], which is used to eliminate selected foreground pixels from binary images. It is generally used for skeletonization. Thinning is usually particularly applied to binary images, and results in another binary image as output. Thinning

Original Image   **Binary Image**   Hit-and-Miss Output

FIGURE 6.8
Output of hit and miss of a binary image.

FIGURE 6.9
Output of thinning of a binary image.

FIGURE 6.10
Output of thickening of a binary image.

operations can be expressed by hit and miss operations. The thinning of an image I by a structuring element S is:

$$\text{Thin}(I,S) = I - \text{hit\_and\_miss}(I,S), \tag{6.10}$$

where "$-$" is a logical subtraction and can be denoted by two binary images A and B: $A - B = A \cap \bar{B}$. Figure 6.9 shows the output of a thinning operation.

6.1.7 Thickening

Thinning operation is to some extent like closing or dilation [8], which is used to grow selected foreground pixels from binary images. It is usually used to determine the estimated convex hull of a shape. Thickening is normally only applied to binary images, and it produces another binary image as output. A thickening operation can be expressed by a hit and miss operation. The thickening of an image I by a structuring element S is:

$$\text{Thick}(I,S) = I \cup \text{hit\_and\_miss}(I,S). \tag{6.11}$$

Thus, the thickened image consists of the original image and additional foreground pixels produced by the hit-and-miss operation. Figure 6.10 shows the output of a thickening operation.

6.2 Grayscale Morphology

Grayscale morphology [9] is basically a multidimensional simplification of the binary morphology.

Original Image

Erosion Output

FIGURE 6.11
Output of erosion of a grayscale image.

6.2.1 Erosion

The erosion of a grayscale image $I(x, y)$ by the structuring element $S(x, y)$ can be expressed as:

$$I \ominus S = \min_{s,t \in B}\{I(x+s, y+t) - S(x, y)\}. \tag{6.12}$$

Here, B is the space in which $S(x, y)$ is defined. In grayscale erosion, the local minimum grey level in the image is considered over the region defined by the structuring element [10]. The image becomes darker with the erosion operation and light details are reduced. Figure 6.11 shows the output of an erosion operation.

6.2.2 Dilation

The dilation of a grayscale image $I(x, y)$ by the structuring element $S(x, y)$ can be expressed as:

$$I \oplus S = \max_{s,t \in B}\{I(x-s, y-t) + S(x, y)\}. \tag{6.13}$$

Here, B is the space in which $S(x, y)$ is defined. In grayscale dilation, the local maximum gray level in the image is considered over the region defined by the structuring element [11]. The image become brighter by the dilation operation and dark details are reduced. Figure 6.12 shows the output of a dilation operation.

6.2.3 Opening

The opening of a grayscale image $I(x, y)$ by the structuring element $S(x, y)$ can be expressed as:

$$I \circ S = (I \ominus S) \oplus S. \tag{6.14}$$

Original Image   **Dilation Output**

FIGURE 6.12
Output of dilation of a grayscale image.

Original Image   **Opening Output**

FIGURE 6.13
Output of opening of a grayscale image.

In the grayscale opening, bright details are reduced. Figure 6.13 shows the output of the opening operation.

6.2.4 Closing

The closing of a grayscale image $I(x, y)$ by the structuring element $S(x, y)$ can be expressed as:

$$I \cdot S = (I \oplus S) \ominus S. \tag{6.15}$$

In grayscale closing, the dark details are reduced. Figure 6.14 shows the output of the closing operation.

6.3 Summary

Morphology is used to remove the imperfections of the image. The output of morphology is generated by using a structuring element. Morphology

Original Image

Closing Output

FIGURE 6.14
Output of closing of a grayscale image.

operations are applicable in binary as well as grayscale images. Some examples of morphological operations are *erosion*, which is used to shrink binary images and darken grayscale images; *dilation*, which is used to fill up gaps in binary images and lighten grayscale images; *opening*, which is used to open up gaps between connected regions within binary images and where bright details of grayscale images are reduced; *closing*, which is used to connect or fill holes in binary image regions and where dark details of grayscale images are reduced; *thinning*, which is mainly applied to binary image for thinning and thickening, and is used to determine the estimated convex hull of a binary shape.

References

1. Kotyk, T., Ashour, A. S., Chakraborty, S., Dey, N., & Balas, V. E. 2015. Apoptosis analysis in classification paradigm: A neural network based approach. In *Healthy World Conference—A Healthy World for a Happy Life* (pp. 17–22). Kakinada (AP), India.
2. Shih, F. Y. 2009. *Image Processing and Mathematical Morphology: Fundamentals and Applications*. CRC Press, Boca Raton, FL.
3. Chaki, J., Parekh, R., & Bhattacharya, S. 2016, January. Plant leaf recognition using a layered approach. In *Microelectronics, Computing and Communications (MicroCom), 2016 International Conference on* (pp. 1–6). IEEE, Durgapur, India.
4. Chaki, J., Parekh, R., & Bhattacharya, S. In press. Plant leaf classification using multiple descriptors: A hierarchical approach. *Journal of King Saud University-Computer and Information Sciences*, doi:10.1016/j.jksuci.2018.01.007.
5. Ravi, S., & Khan, A. M. 2013. Morphological operations for image processing: Understanding and its applications. In *NCVSComs-13 Conference Proceedings* (pp. 17–19).
6. Pardeshi, R., Chaudhuri, B. B., Hangarge, M., & Santosh, K. C. 2014, September. Automatic handwritten Indian scripts identification. In *Frontiers in Handwriting*

Recognition (ICFHR), 2014 14th International Conference on (pp. 375–380). IEEE, Heraklion, Greece.

7. Benavent, X., Dura, E., Vegara, F., & Domingo, J. 2012. Mathematical morphology for color images: An image-dependent approach. *Mathematical Problems in Engineering*, 2012(678326), 1–18.

8. Sonka, M., Hlavac, V., & Boyle, R. 2014. *Image Processing, Analysis, and Machine Vision*. Cengage Learning, Stamford, USA.

9. Sternberg, S. R. 1986. Grayscale morphology. *Computer Vision, Graphics, and Image Processing*, 35(3), 333–355.

10. Ćurić, V., Landström, A., Thurley, M. J., & Hendriks, C. L. L. 2014. Adaptive mathematical morphology—A survey of the field. *Pattern Recognition Letters*, 47, 18–28.

11. Wang, Y., Shi, F., Cao, L., Dey, N., Wu, Q., Ashour, A. S., & Wu, L. In press. Morphological segmentation analysis and texture-based support vector machines classification on mice liver fibrosis microscopic images. *Current Bioinformatics*.

7

Other Applications of Image Preprocessing

7.1 Preprocessing of Color Images

A color image has a huge quantity of information. If the color information is hidden, human eyes may fail to analyze it [1]. Moreover, small alterations in image features such as color, intensity, texture, and so forth are truly hard to accomplish. Thus, preprocessing of color images is needed to preserve the reliability of edges and other detailed information needed for further processing.

There are two types of color image processing: pseudo color or false color processing, and full color or true color processing. The purpose of pseudo color processing is to color a grayscale image by assigning different colors in different intensity ranges of a gray level image [2]. Pseudo color is also called false color, as the colors are not originally present in the grayscale image. The human eye can interpret about two dozens of gray shades in a grayscale image, whereas it can interpret nearly 1000 variations of colors in a color image [3]. Thus, if a given grayscale image is converted to color by using pseudo color processing, the interpretation of different intensities becomes much more convenient, as compared to an ordinary grayscale image. Pseudo coloring can be done by an intensity slicing method. Suppose there are L number of intensity values in a grayscale image $I(x, y)$, which varies from $0,...,(L-1)$. In this case, l_0 represents black where $I(x, y) = 0$ and l_{L-1} represents white where $I(x, y) = L - 1$. Suppose there are P number of planes perpendicular to the intensity plane where $0 < P < L - 1$. These planes are placed to the intensity levels $l_1, l_2..., l_P$. P number of planes divide the intensities to $P + 1$ number of intervals. So, the color C_k is assigned to the gray level intensity at position (x, y) can be denoted by $f(x, y) = C_k$ if $I(x, y) \in D_k$ where D_k is the intensity range between l_k and l_{k+1}. Thus, it can be said that P number of planes divide the intensities to $P + 1$ number of intervals denoted by $D_1, D_2,..., D_{P+1}$. By using this concept the gray level intensity range can be divided into some intervals, and for each interval a particular color can be assigned. In this way a grayscale image can be colored—this procedure is known as pseudo coloring. Figure 7.1 shows the pseudo coloring image of a grayscale image [4]. In the pseudo color image we can visualize different

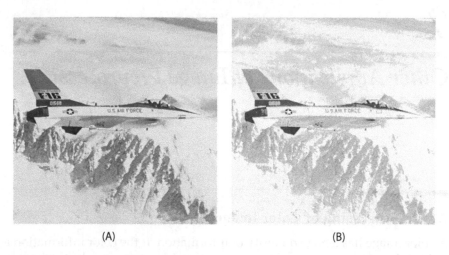

(A)   (B)

FIGURE 7.1
(A) Grayscale image, (B) Pseudo color image.

intensities of the image region with different colors, which are almost flat in the grayscale image. So, using a pseudo color image, intensities of the image are much more interpretable or distinguishable than in a grayscale image. In case of an RGB image, colors are added to R, G, and B channels separately and the combination of R, G, and B channels enables the interpretation of pseudo color images [5].

Grayscale to color image conversion can be done by the transformations shown in Figure 7.2.

In Figure 7.2 $I(x, y)$ is the grayscale image, which is transformed by three different transformations: RED transformation, GREEN transformation, and BLUE transformation [6]. RED, GREEN, and BLUE transformations give the red, green, and blue plane output of the input grayscale image, which is given by $I_R(x, y)$, $I_G(x, y)$, and $I_B(x, y)$. When these three planes are combined together and displayed in a color display system it is known as a pseudo color

FIGURE 7.2
Grayscale to color transformation.

(A) (B)

FIGURE 7.3
(A) Grayscale image, (B) Pseudo color transformed image.

image. For example, Equation 7.1 denotes the transformation functions used to generate the color image, and Figure 7.3 shows the color transformation of a grayscale image by using Equation 7.1:

$$I_R(x,y) = I(x,y)$$
$$I_G(x,y) = 0.33 \times I(x,y)$$
$$I_B(x,y) = 0.11 \times I(x,y).$$

(7.1)

In this example, to convert the grayscale image to color, the exact intensities of the grayscale image are copied to the red plane, but the degraded version of intensities of the original grayscale image are used in the green and blue plane. The combination of this red, green, and blue plane is shown in Figure 7.3.

In the full color, or true color image preprocessing, the actual color of the image is considered [7]. In such types of images, the colors can be specified by using different color models like RGB (Red-Green-Blue), HSI (Hue-Saturation-Intensity), CMY (Cyan-Magenta-Yellow), and so on. In some cases, color image processing can be more convenient in a particular color model while less convenient in some other color model. In such cases, the image is converted from one color model to another color model. Figure 7.4 shows the representation of different color components, or color planes, of an image in the RGB color model.

Figure 7.5 shows the representation of different color components, or color planes, of an image in the HSI color model.

Figure 7.6 shows the representation of different color components, or color planes, of an image in the CMY color model.

FIGURE 7.4
Red, Green, and Blue plane of RGB color image.

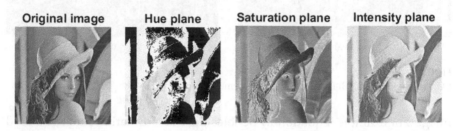

FIGURE 7.5
Hue, Saturation and Intensity plane of HSI color image.

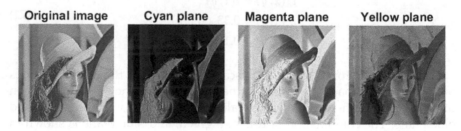

FIGURE 7.6
Cyan, Magenta, and Yellow plane of CMY color image.

Different preprocessing transformation [8] operations can be done in these color models, such as intensity modification represented by Equation 7.2:

$$f(x,y) = T \times O(x,y), \qquad (7.2)$$

where $0 < T < 1$, $O(x, y)$ is the input image and $f(x, y)$ is the processed image. So, if the image in the RGB color space is considered, then Equation 7.2 can be rewritten, as shown in Equation 7.3:

$$f_i(x,y) = T \times O_i(x,y), \qquad (7.3)$$

where $i = 1,2$, and 3, which represents the red, green, and blue planes of the RGB color model, respectively. From Equation 7.3 it can be said that in case of an RGB color image, all the three planes are scaled by the same scaling factor, T.

If intensity modification is done for an HSI color image, then the scaling is done only in the I plane of the input image, as this is the only plane of the HSI color model representing the intensity. The hue and saturation of the processed image will remain the same as in the input image. Thus, in this case Equation 7.2 can be rewritten, as shown in Equation 7.4:

$$f_1(x,y) = O_1(x,y)$$
$$f_2(x,y) = O_2(x,y) \quad\quad (7.4)$$
$$f_3(x,y) = T \times O_3(x,y).$$

Similarly, the intensity modification in CMY color space can be represented by Equation 7.5:

$$f_i(x,y) = T \times O_i(x,y) + (1-T), \quad\quad (7.5)$$

where $i = 1,2$, and 3, which represents the cyan, magenta, and yellow planes of the CMY color model, respectively.

From these intensity transformations it can be said that the computation using the HSI color space is minimum, as compared to RGB and CMY color space, because in the RGB and CMY color space the scaling is done in all three color planes. Figure 7.7 shows the intensity-modified output for the HSI color image with a scaling factor of 0.5.

Color complement is another preprocessing [9] transformation, which can be done in true color image. Let us first take a look at the color wheel, or color circle, which is shown in Figure 7.8.

In Figure 7.8, it can be seen that the colors diagonally opposite in the color wheel are complementary colors, such as cyan which is the complementary

(A)   (B)

FIGURE 7.7
(A) Original Image, (B) Intensity modified image.

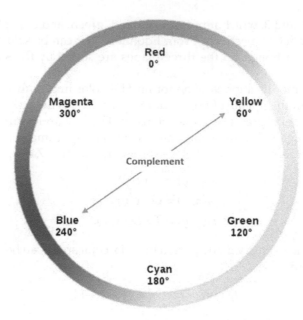

FIGURE 7.8
Color wheel.

color of red and vice versa, or yellow which is the complementary color of blue and vice versa, and so on. Color complement is analogous to grayscale negative. Thus, if the same transform function, which is used to generate grayscale negative, is applied in different planes of the color image, a color complement image is generated. This can be represented by Equation 7.6 and Figure 7.9, which shows the color complement image of an RGB color input image:

$$f_i(x,y) = L - 1 - O_i(x,y), \tag{7.6}$$

where $i = 1,2$, and 3, $O_i(x, y)$ is the input image, L is the maximum number of color shades (in case of RGB $L = 256$), and $f_i(x, y)$ is the processed image.

From Figure 7.9 it can be seen that the complement of an input image looks like the photographic negative of a color image.

Color slicing is the next preprocessing [10] transformation that can be applied to the true color or full color image. Color slicing is used to highlight a certain color range in an image, and thus it is useful to find an object of a certain color in an image. In this method, it is assumed that all the colors of interest lie within a cube of width, say W, centered at the prototypical color whose components are given by some vector say C_1, C_2, and C_3, as shown in Figure 7.10.

Original image

Complementary image

FIGURE 7.9
Complementary image output.

The color slicing transformation can be denoted by Equation 7.7:

$$f_i(x,y) = \begin{cases} 0.5 \text{ if } \left[|O_i(x,y) - C_i| > \dfrac{W}{2} \right] \forall 1 \le i \le 3. \\ O_i(x,y), \qquad\qquad\qquad \text{otherwise} \end{cases} \tag{7.7}$$

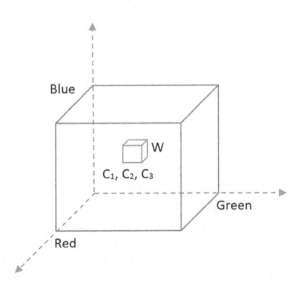

FIGURE 7.10
Cube of width W.

Original image

Color sliced image

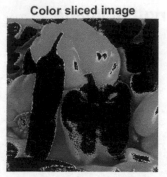

FIGURE 7.11
Output of color slicing.

This means all the colors outside the cube of width W will be represented by some insignificant color, but inside the cube the original colors are retained. Figure 7.11 shows the output of the color slicing where only the red shades are kept.

The next type of preprocessing transformation of the color image is tone correction [11]. This is again analogous with the intensity enhancement, or contrast enhancement, of the grayscale image. A color image may have a flat tone, light tone, or dark tone. These tones represent the distribution of different color intensities within the color image or RGB image. The form of transformation function used to correct the tone of flat, light, and bark-toned image is shown in Figure 7.12.

In case of light tone image wide range of intensities in the input image is mapped to narrow range intensities in the output image so that the output image become dark. In case of dark tone image narrow range of intensities in the input image is mapped to wide range intensities in the output image so that the output image become light.

Other types of color image preprocessing involve histogram equalization, segmentation of color images, and so on Figure 7.13 through 7.16.

7.2 Image Preprocessing for Neural Networks and Deep Learning

Deep learning has really become a main research area in the past few years [12]. Deep learning uses neural networks, which need a large number of training data and are comprised of many of hidden layers. These models are used in speech, vision, image, video, language processing, and so forth.

For an image, providing the image pixel values directly into a neural network may cause numerical overflows [13,14]. Also, some objective and activation

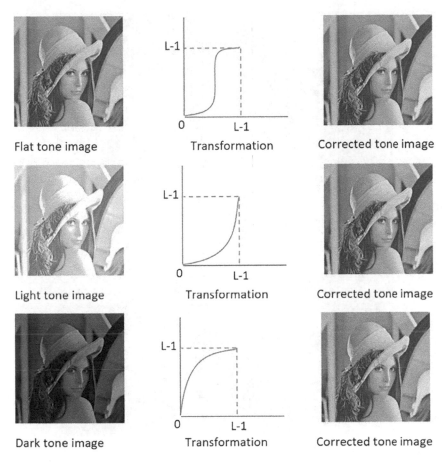

FIGURE 7.12
Tone correction.

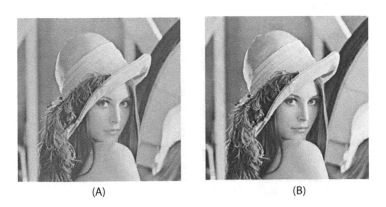

FIGURE 7.13
(A) Original image, (B) Image output after histogram equalization.

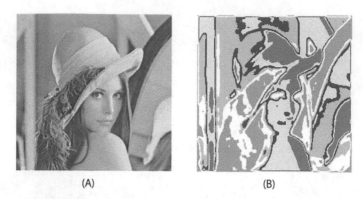

(A)   (B)

FIGURE 7.14
(A) Original image, (B) Segmented output using 5 bins.

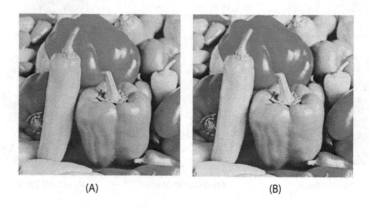

(A)   (B)

FIGURE 7.15
(A) Original Image, (B) Image output after filtering or masking. Here only the red shades are kept in color and rest of the image is desaturated.

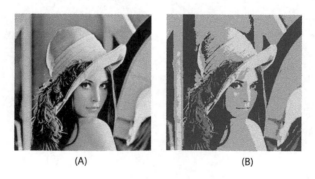

(A)   (B)

FIGURE 7.16
(A) Original image, (B) Image Otsu thresholding output.

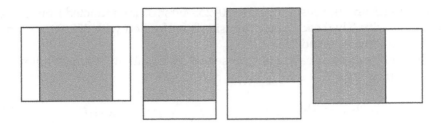

FIGURE 7.17
Cropping of image data.

functions are not compatible with all kinds of input. The wrong arrangement produces a poor result during the learning phase of a neural network [15–17]. To construct an efficient neural network model, cautious attention is required to build the network architecture as well as the input data format. The most common image data input factors are the number of images, image width, image height, number of levels per pixel, and number of channels. For an RGB image, there are three channels of data representing the colors (pixel intensity values) in Red, Green, and Blue channels, which range between 0 and 255 [18,19].

A number of preprocessing steps are needed prior to using this in any Deep Learning project. Some of the most common preprocessing steps are discussed below.

Unvarying Aspect Ratio: Most of the neural networks presume that the input image is square in shape. So, it is essential to check every image to ensure it is square or not [20], and cropped properly, as shown in Figure 7.17. While cropping, usually the center part is kept.

Scaling of Images: After making all images square in shape, scaling of each image is properly done [21,22]. For example, suppose the image is of size 250×250 pixels and we have to obtain an image with a height and width of 100 pixels. Therefore, the height and width of each image are scaled by a factor of 0.4 (100/250). The same applies for up-scaling.

Normalization of Image Inputs: Image data normalization [23,24] is a vital step, which confirms a similar data distribution for every input data. This helps to converge the network faster while training it. In image processing, normalization helps to change the pixel intensity range. There are three types of normalization: data rescaling, data standardization, and data stretching. Data rescaling is further divided into linear and nonlinear rescaling.

The linear data scaling can be represented by Equation 7.8:

$$I_{\text{Norm}} = (I - I_{\text{Min}})\frac{I_{\text{NewMax}} - I_{\text{NewMin}}}{I_{\text{Max}} - I_{\text{Min}}} + I_{\text{NewMin}}, \tag{7.8}$$

where I_{Max} and I_{Min} are the maximum and minimum intensities of the original image, and I_{NewMax} and I_{NewMin} are the maximum and minimum intensities of the normalized image. For example, suppose the image has the intensity range 30–120 and the desired range is 0–255. First, 30 is subtracted from every pixel intensity. Then each pixel intensity is multiplied with 255/90, making the range between 0 and 255.

The nonlinear data scaling is represented by Equation 7.9, which follows a sigmoid function:

$$I_{\text{Norm}} = (I_{\text{NewMax}} - I_{\text{NewMin}}) \frac{1}{1 + e^{-((1-\beta)/\alpha)}} + I_{\text{NewMin}}, \tag{7.9}$$

where β denotes the intensity around which the range is centered, and α denotes the width of the input intensity.

Data standardization is the second way to normalize image data, where the average of the data is subtracted from the image and divided by its standard deviation. The spreading of such data looks like a Gaussian curve with mean = 0, and a standard deviation (std) = 1. Data standardization can be represented by Equation 7.10:

$$I_{\text{Norm}} = \frac{I - I_{\text{Mean}}}{I_{\text{Std}}}. \tag{7.10}$$

Data stretching is the third way to normalize image data, where the data are braced to a maximum and minimum value, and can be represented by using Equation 7.11:

$$\begin{aligned} I_{\text{Norm}}[I < c] &= c \\ I_{\text{Norm}}[I > d] &= d. \end{aligned} \tag{7.11}$$

Here, image data values greater than d are set to d, and the same occurs inversely with c.

Reduction in Dimension: Sometimes the three channels of an RGB image [25] are collapsed into a single grayscale channel. Reduction in the dimension of image data is often needed when the neural network performance is permitted to be dimension-invariant.

Augmentation of Image Data: The next preprocessing technique [26] includes augmenting the image data with disturbed versions of the present images. Rotation, scaling, and other affine transformations are usually used to augment image data. This prevents the neural network from recognizing unwanted characteristics present in the disturbed version of image data.

7.3 Summary

The need of preprocessing of color images in the field of Deep Learning is discussed in this chapter. Color image processing includes pseudo color and full color or true color processing. The purpose of pseudo color processing is to color a grayscale image by assigning different colors to different intensity ranges of a gray level image. In the case of an RGB image, colors are added to the R, G, and B channels separately, and the combination of R, G, and B channels allows for the interpretation of a pseudo color image. Through pseudo color images, we can visualize different intensities of the image region with a different color, which would be almost flat in the grayscale image. Thus, using the pseudo color image, intensities of the image are much more interpretable or distinguishable than for a grayscale image. In the full-color image, the actual color of the image is considered. In such types of images, the colors can be specified by using different color models like RGB (Red-Green-Blue), HSI (Hue-Saturation-Intensity), CMY (Cyan-Magenta-Yellow), and so on. Different preprocessing transformation operations can be performed on these color models such as intensity modification, color complement, color slicing, tone correction, histogram equalization, segmentation of the color image, and so forth.

References

1. Ghosh, A., Sarkar, A., Ashour, A. S., Balas-Timar, D., Dey, N., & Balas, V. E. 2015. Grid color moment features in glaucoma classification. *Int J Adv Comput Sci Appl*, 6(9), 1–14.
2. Dey, N., Ashour, A. S., Chakraborty, S., Samanta, S., Sifaki-Pistolla, D., Ashour, A. S., & Nguyen, G. N. 2016. Healthy and unhealthy rat hippocampus cells classification: A neural based automated system for Alzheimer disease classification. *Journal of Advanced Microscopy Research*, 11(1), 1–10.
3. Chaki, J., Parekh, R., & Bhattacharya, S. 2017, March. An efficient fragmented plant leaf classification using color edge directivity descriptor. In *International Conference on Computational Intelligence, Communications, and Business Analytics* (pp. 197–211). Springer, Singapore.
4. Li, Z., Shi, K., Dey, N., Ashour, A. S., Wang, D., Balas, V. E., ... & Shi, F. 2017. Rule-based back propagation neural networks for various precision rough set presented KANSEI knowledge prediction: A case study on shoe product form features extraction. *Neural Computing and Applications*, 28(3), 613–630.
5. Bhattacharya, T., Dey, N., & Chaudhuri, S. R. 2012. A session based multiple image hiding technique using DWT and DCT. *arXiv preprint arXiv:1208.0950*.
6. Candemir, S., Borovikov, E., Santosh, K. C., Antani, S., & Thoma, G. 2015. Rsilc: Rotation-and scale-invariant, line-based color-aware descriptor. *Image and Vision Computing*, 42, 1–12.
7. Chaki, J., & Parekh, R. 2011. Plant leaf recognition using shape based features and neural network classifiers. *International Journal of Advanced Computer Science and Applications*, 2(10) 41–47.
8. Benavent, X., Dura, E., Vegara, F., & Domingo, J. 2012. Mathematical morphology for color images: An image-dependent approach. *Mathematical Problems in Engineering*, 2012(678326) 1–18.

9. Sonka, M., Hlavac, V., & Boyle, R. 2014. *Image Processing, Analysis, and Machine Vision*. Cengage Learning, Stamford, USA.

10. Fu, K. S. 2018. *Special Computer Architectures for Pattern Processing*. CRC Press, Boca Raton, FL.

11. Kotyk, T., Ashour, A. S., Chakraborty, S., Dey, N., & Balas, V. E. 2015. Apoptosis analysis in classification paradigm: A neural network based approach. In *Healthy World Conference—A Healthy World for a Happy Life* (pp. 17–22). Kakinada (AP), India.

12. Dong, C., Loy, C. C., He, K., & Tang, X. 2016. Image super-resolution using deep convolutional networks. *IEEE Transactions on Pattern Analysis and Machine Intelligence*, 38(2), 295–307.

13. Nimmy, S. F., Sarowar, M. G., Dey, N., Ashour, A. S., & Santosh, K. C. 2018. Investigation of DNA discontinuity for detecting tuberculosis. *Journal of Ambient Intelligence and Humanized Computing*, 1–15.

14. Chaki, J., Parekh, R., & Bhattacharya, S. 2015. Plant leaf recognition using texture and shape features with neural classifiers. *Pattern Recognition Letters*, 58, 61–68.

15. Li, Z., Dey, N., Ashour, A. S., Cao, L., Wang, Y., Wang, D., & Shi, F. 2017. Convolutional neural network based clustering and manifold learning method for diabetic plantar pressure imaging dataset. *Journal of Medical Imaging and Health Informatics*, 7(3), 639–652.

16. Chaki, J., Parekh, R., & Bhattacharya, S. In press. Plant leaf classification using multiple descriptors: A hierarchical approach. *Journal of King Saud University-Computer and Information Sciences*, doi:10.1016/j.jksuci.2018.01.007.

17. Halder, C., Obaidullah, S. M., Santosh, K. C., & Roy, K. 2018. Content independent writer identification on Bangla script: A document level approach. *International Journal of Pattern Recognition and Artificial Intelligence*, 32(9), 1856011.

18. Chatterjee, S., Sarkar, S., Hore, S., Dey, N., Ashour, A. S., Shi, F., & Le, D. N. 2017. Structural failure classification for reinforced concrete buildings using trained neural network based multi-objective genetic algorithm. *Structural Engineering and Mechanics*, 63(4), 429–438.

19. Chaki, J., Parekh, R., & Bhattacharya, S. 2016, January. Plant leaf recognition using a layered approach. In *Microelectronics, Computing and Communications (MicroCom), 2016 International Conference on* (pp. 1–6). IEEE.

20. Chatterjee, S., Hore, S., Dey, N., Chakraborty, S., & Ashour, A. S. 2017. Dengue fever classification using gene expression data: A PSO based artificial neural network approach. In *Proceedings of the 5th International Conference on Frontiers in Intelligent Computing: Theory and Applications* (pp. 331–341). Springer, Singapore.

21. Chaki, J., Parekh, R., & Bhattacharya, S. 2015, July. Recognition of whole and deformed plant leaves using statistical shape features and neuro-fuzzy classifier. In *Recent Trends in Information Systems (ReTIS), 2015 IEEE 2nd International Conference on* (pp. 189–194). IEEE.

22. Samanta, S., Ahmed, S. S., Salem, M. A. M. M., Nath, S. S., Dey, N., & Chowdhury, S. S. 2015. Haralick features based automated glaucoma classification using back propagation neural network. In *Proceedings of the 3rd International Conference on Frontiers of Intelligent Computing: Theory and Applications (FICTA) 2014* (pp. 351–358). Springer, Cham.

23. Santosh, K. C., & Nattee, C. 2007. Template-based nepali natural handwritten alphanumeric character recognition. *Science & Technology Asia*, 12(1), 20–30.

24. Hore, S., Chatterjee, S., Sarkar, S., Dey, N., Ashour, A. S., Balas-Timar, D., & Balas, V. E. 2016. Neural-based prediction of structural failure of multistoried RC buildings. *Structural Engineering and Mechanics*, 58(3), 459–473.
25. Maji, P., Chatterjee, S., Chakraborty, S., Kausar, N., Samanta, S., & Dey, N. 2015, March. Effect of Euler number as a feature in gender recognition system from offline handwritten signature using neural networks. In *Computing for Sustainable Global Development (INDIACom), 2015 2nd International Conference on* (pp. 1869–1873). IEEE.
26. Bhattacherjee, A., Roy, S., Paul, S., Roy, P., Kausar, N., & Dey, N. 2016. Classification approach for breast cancer detection using back propagation neural network: A study. In *Biomedical Image Analysis and Mining Techniques for Improved Health Outcomes* (pp. 210–221). IGI Global, Hershey, Pennsylvania.

Index

9 780367 570804